Project Management Recipes for Success

Project Management Recipes for Success

Guy L. De Furia

CRC Press
Taylor & Francis Group
Boca Raton London New York

CRC Press is an imprint of the
Taylor & Francis Group, an **informa** business

AN AUERBACH BOOK

CRC Press
Taylor & Francis Group
6000 Broken Sound Parkway NW, Suite 300
Boca Raton, FL 33487-2742

First issued in hardback 2017

© 2009 by Guy L. De Furia
CRC Press is an imprint of Taylor & Francis Group, an Informa business

No claim to original U.S. Government works

ISBN-13: 978-1-4200-7824-4 (pbk)
ISBN-13: 978-1-138-44043-2 (hbk)

Library of Congress Cataloging-in-Publication Data

De Furia, Guy L.
   Project management recipes for success / Guy L. De Furia.
      p. cm. -- (ESI international project management series ; 6)
   Includes bibliographical references and index.
   ISBN 978-1-4200-7824-4 (alk. paper)
   1. Project management. I. Title. II. Series.

HD69.P75D437 2009
658.4'04--dc22   2008025581

Visit the Taylor & Francis Web site at
http://www.taylorandfrancis.com

and the CRC Press Web site at
http://www.crcpress.com

Dedication

I would like to dedicate this book to four people who helped me more than they realized when I was a boy: my grandmother Maria Giordino, who thought I was special; my uncles Carmen Giordino and John Caruso, who showed me by how they lived, what it means to be a man; and especially my wife Barbara, my Polish girl, who continues to be one leg of the compass around which my life revolves.

Contents

Contents

List of Figures

List of Tables

List of Equations

Preface

Project management has existed since the first time humans worked together to achieve some common purpose. Perhaps it was to build a shelter or secure food. Clearly, any effort that required the combined effort of many people over long periods of time would require some project management technology. History provides many examples — some ancient and some recent. The building of the pyramids about four thousand years ago comes to mind as well as the NASA projects to reach the moon. Until relatively recent times, project management has been a subset of other disciplines: civil engineering, shipbuilding, etc. Today, project management is a discipline in and of itself. Project management is a technology with its own set of principles, procedures, and tools. The purpose of this book is to explain how to use this technology.

Recipes for Success

One of the things I like to do is cook. Now I'm not a great scholar of the culinary art and science but I do know how to follow a recipe. I have been significantly aided in my "forays" into my wife's kitchen (yes, it is HER kitchen!) by her extensive library of cookbooks. My strategy is to find a recipe for a meal that seems inviting, collect the ingredients, and then follow the instructions printed in the book. Over the years I have repeated some recipes many times. For these meals, I have gained experience so that I now modify the recipe to my own desires. The first few times I used a recipe, I followed it exactly and noted the outcome of each step. At the beginning, I did not know why the ingredients and procedures worked, only that they produced the desired product. As I gained confidence, I used the experience (called lessons learned) to modify the recipe. My early successes encouraged me to explore deeper into the art and technology of cooking. I see managing a project as analogous to cooking a meal.

What do beginning project managers need? I think they need a single reference that provides the most basic concepts, tools, and procedures with which to initiate, plan, execute, and close a project to success, just as beginning chefs need a cookbook with which to achieve early cooking successes! The compact disk that accompanies this book includes all the tools, format outlines, and templates discussed in this book.

Purpose of This Book

The purpose of this book is to provide the ingredients and procedures (call them the recipes) for achieving a successful project. We'll start out by listing the ingredients and procedures for running a successful project, and conclude by explaining some of the finer points of managing and controlling the project. Because this book is a cookbook, it will not discuss theoretical issues nor will it devote a lot of space to why the procedures work. The procedures in project management are relatively simple to accomplish; project difficulties come not from doing them poorly but from not doing them at all.

Audience

This book is written for three classes of people: those responsible for the hands-on planning and managing of projects, program managers who oversee project managers, and senior executives responsible for providing the guidance and financial support to project and program managers. Each will get something from this book. Project managers will gain the confidence that comes from following a good recipe for successful projects. Program managers will gain a perspective on the myriad activities their project managers are performing to achieve a well-disciplined project. Senior managers will gain a perspective of the disciplined approach necessary at the beginning of a project to reduce the number of ill-advised projects and the total effort required to achieve successful projects.

Overview of Contents

Chapter 1 provides an overview of project management and the typical project life cycle. It introduces the powerful political dynamics associated with managing a project. It also puts into perspective the issues associated with differing levels of planning definitiveness and sophistication as well as the value of people versus hard skills.

Chapter 2 discusses the work required before a project is approved for planning and execution. It outlines the activities that should occur during the initiation phase: needs identification, project definition, determination of financial viability, and project approval. It concludes with the difficulties that may arise when the activities appropriate to the initiation phase are not done. This chapter discusses the various ways organizations may select a project for funding from a list of alternative project proposals. It highlights some of the difficulties experienced by projects that have inadequate or incomplete formalization before project launch. It includes an extensive discussion of various ways organizations may select a project for funding from a list of alternative projects proposals. It also highlights some of the difficulties experienced by projects that have inadequate or incomplete formalization before project launch. This chapter should be particularly useful to senior managers responsible for establishing the ground rules and culture in which projects may thrive or wither.

Chapter 3 discusses the extensive list of activities that should occur during the planning phase as well as the order in which they should be done. It makes the distinction between the project plan and the project baseline and it provides recommendations on what planning documents should be included in the baseline. This chapter lists the project manager's responsibilities as well as the difficulties that may arise during this phase.

Chapter 4 describes how to prepare the work breakdown structure (WBS) and the work package work order. These documents provide definitive information about the scope of the project.

Chapter 5 describes how to estimate work package cost and determine the project total cost. It includes a discussion of the various elements of cost and how they are included in budgeting for the project.

Chapter 6 deals with the time dimension. It describes how to generate the documents (precedent diagram, Gantt chart, and milestone chart) with which to determine and control the project schedule.

Chapter 7 describes the process for developing the project resource plan. With this plan, the project manager coordinates the implementation activities and avoids delays and cost increases caused by such things as work package managers not showing up at the right place at the right time, materials not being available when needed, etc.

Chapter 8 describes the need for a project filing system to deal with the myriad pieces of paper and electronic data generated by a project.

Chapter 9 discusses risk management and the procedures for developing viable risk management and opportunity management plans. It provides an extensive discussion of risk with tools for dealing with common types of risk management decisions.

Chapter 10 describes the project baseline and includes procedures for establishing the measurement system with which to control cost and time.

Chapter 11 describes how to establish a scope change system. It includes templates and forms for this purpose.

Chapter 12 discusses the activities associated with the close of a project. It discusses the benefits of a well-executed project close as well as the activities that should be included. It includes a list of suggested closeout phase outputs as well as the difficulties that may occur at the end of a project

Chapter 13 discusses the activities associated with the implementation phase. An explanation of earned value methodology is included; it is used to measure current performance and to predict project final cost and final completion date. Keeping cost, schedule, and scope variance within acceptable limits is extensively covered as well as problems associated with keeping a project "on track."

Chapter 14 outlines the issues associated with the project closeout phase and the activities that should be performed to end a project in a disciplined, professional way.

Acknowledgments

I would like to thank my wife of many years for having the faith in me to keep me during the lean years. She has generously shared her faith, ideas, and good cheer with me even when I may not have deserved these. Sometimes I think that she continues to feed and care for me for reasons well beyond love. Perhaps she considers me one of her numerous stray cats and dogs that she takes in and cares for every day. For whatever reason, I am deeply grateful and much more for her support. I would like to thank her for her indulgence in letting me mess up HER kitchen whenever the urge to cook overpowers me. I want to thank my father because he taught me much more than he realized. It is from my time in his company that I learned the love of work. I would also like to thank my children Paul, Allison, Ash, and Michael, and my granddaughter Mikaela and grandson Alex for their understanding of my efforts and their love. In many ways, the time spent on this book has been at the expense of time with them.

I would like to thank the people at the Educational Services Institute (ESI) International, especially Sally Spooner, Diane Davenport, and Keely Niebrzydowski, who allowed me to teach a broad range of project management courses. It is the experience from teaching these courses that gave me the confidence to write this book. It was the interaction with my students that convinced me that there is a need for another book on project management. Finally, I would also like to thank the editor of the ESI Web *Horizon* newsletter, Jessica McCaughley, for her encouragement. It was her feedback that convinced me that I had something useful to say and could do so in a simple way.

About the Author

Dr. Guy L. De Furia is a senior faculty member of the Educational Services Institute (ESI) International. He has over thirty years of experience as an organizational development consultant and educator. He has lectured on how to achieve successful organizational development consulting, the management of change, organizational structuring, horizontal systems interventions, and the training specialist as an agent of change and project management.

Dr. De Furia spent seven years as a senior organizational consultant to the Department of the Army. In this position, he conducted assessment, design, implementation, and evaluation of strategic planning, organizational development, performance management, counseling, communications, and

Photograph by Radostina Stoycheva

team-building interventions and workshops. He is the author of three assessment instruments and a book on interpersonal trust. Dr. De Furia was project leader and principal author of the U.S. Marine Corps *Manpower Requirements Determination Manual.* He is also the author of many courses including those intended to teach manpower estimating techniques and statistical regression forecasting. While at the Military District of Washington, he wrote the *Handbook of Cost Benefit Analysis* to assist managers in determining the costs and benefits associated with government facilities versus those contracted resources.

While at Hay Systems, Inc., Dr. De Furia was director of Human Resources Consulting, responsible for the analysis, design, development, implementation, and evaluation of executive-level training and organization development efforts. While with the U.S. Army Electronics Command, Dr. De Furia performed methods improvement (organizational re-engineering) studies and led a statistical analysis team. He established quantitative budget analysis and also developed and taught linear regression techniques.

While at Fyr-Fyter Corporation, he was a project engineer in charge of design, construction, and installation of municipal fire alarm systems.

Dr. De Furia received engineer technician certificates in mechanical and electro-mechanical engineering from Newark College of Engineering. He received a bachelor's degree in psychology from Rutgers University, and a master's degree in general experimental psychology from Fairleigh Dickinson University. He did his doctoral studies at Stevens Institute of Technology and completed a doctoral dissertation at St. John's University in management science and organization psychology. He was a faculty member of Stevens Institute of Technology and St. John's University. In addition, he has received the Army Meritorious Civilian Service Award and the Army General Staff Medal for his years of consulting with the Department of Defense. Dr. De Furia is certified as a Project Management Professional (PMP®) by the Project Management Institute (PMI®).

Dr. De Furia is the author of the following articles all of which were published in ESI *Horizons* and available online at www.esi-horizons.com: Measure Schedule Risk Using the Standard Deviation (January 2002), How to Set and Use Project Control Limits (March 2006), Setting and Using Project Control Limits (July 2006), Adding Risk into Project Estimates — PERT vs. Monte Carlo (April 2007), and How to Estimate Risk Probabilities (November 2007).

Chapter 1

Project Management Overview

What Is a Project?

A project is a temporary effort to produce a predetermined outcome. Under the best of conditions, the need for the project will have been clearly established, the definition of the project will have been discussed and resolved, and the economic viability of the project will have been estimated. Economic viability may mean profit; it may mean cost savings; it may mean enhanced productivity. However couched, most projects seek to produce some outcome that solves a problem, and the solution is seen as having economic utility. (Of course, there are exceptions: projects that are funded for purely aesthetic reasons; a work of art is an example.)

All projects have a beginning and an end. The duration may be relatively short as in the moving of an office to a new location, or it may involve many years as in the design and construction of an aircraft carrier.

All projects seek to produce a unique outcome even though the outcome may be similar to previous projects. The building of a house may be similar to houses previously constructed; however, each house has some elements of uniqueness. Hence, every project to build a house has some uniqueness.

All projects involve the interrelated activities that must be completed in an orderly sequence; often many different people perform these activities. Some of these people work for the organization performing the project, and others are hired on a temporary basis to perform specific pieces of the project.

Some projects are small, such as remodeling a kitchen; these are relatively simple in scope, low in budget, and short in duration. Other projects are very large in scope, have large budgets, and take more time to accomplish. The futuristic tangle

1

of highways, interchanges, on- and off-ramps south of Washington, D.C. (sometimes called the "mixing bowl") is such a project. It has been about ten years in the making and it is just nearing completion (we think!).

Project Management

Project management is that part of general business management that specializes in the conditions, procedures, and problems associated with running a project. Some of the procedures in project management are similar to general management (e.g., cost estimating) and others are unique to project management (e.g., network diagramming).

Project Dimensions

All projects have three basic dimensions:

1. Cost to perform the project
2. Time to perform the project
3. Scope of the project

Scope refers to the breadth of work that must be performed to produce all the project deliverables. Some projects have quality as an additional dimension. The basic dimensions of cost, time, and scope must be planned and controlled.

Typical Project Life Cycle

The project life cycle can be described in many ways. Many organizations have identified project life cycles unique to their organization and industry. The generic project life cycle has four phases:

1. Initiation or Concept Phase
2. Planning or Development Phase
3. Execution or Implementation Phase
4. Closeout or Termination Phase

The project life cycle ends when the final deliverable is submitted to and accepted by the customer. This book will use the four-phase model as described here.

Project Initiation

The Initiation Phase should produce three outcomes:

1. A comprehensive understanding of the needs necessitating the project
2. A comprehensive definition of the potential project
3. A committed and informed decision by upper management to proceed with the project

The purpose of a project or the conditions or needs that it will resolve should not be discovered during the Planning or Execution phases; these are articulated before the full definition of the project can be completed. Unfortunately, many projects are launched without this complete understanding of the needs they are to address. It is for this reason that many projects are terminated before completion. An ill-advised project may not discover its purpose if it didn't have one at the beginning. Clarifying a project's needs during the Execution Phase may save the project but it will do so at great cost, in time, money, and human energy.

The third outcome of the Initiation Phase calls for an informed decision to proceed with the project or a decision to invest in another project. To do this, upper management needs information about the potential project.

The second activity of the Initiation Phase calls for a comprehensive definition of the potential project. This project definition is sometimes called the project charter because in some organizations it is the approval of the definition that constitutes the authority to launch the project. The purposes of the project definition are (1) to assure that upper management is fully aware of the conditions, circumstances, and potential risks inherent in the project before it is approved; and (2) to assure that upper management and the project team have the same vision of the project. How to write the project definition is discussed in Chapter 2.

Project Planning

The purpose of the Planning Phase is to

1. Generate the planning documents with which the project will be implemented
2. Develop a cohesive and committed team
3. Establish the professional-trusting relationships with the project's sponsor, customer, stakeholders, and constituencies

Developing the project's planning documents is discussed in Chapters 4 through 12.

Project Implementation

The purpose of the Implementation Phase is to produce the project deliverables. The activities needed to accomplish this include

1. Delegating work
2. Procuring materials and services
3. Controlling the scope of the project
4. Monitoring risk events
5. Monitoring and controlling the project budget

6. Monitoring and controlling the project schedule
7. Forecasting final project cost and delivery date
8. Finding ways to make up time or reduce final cost

These topics are discussed in Chapter 13.

Project Closeout

The Closeout Phase is the time to conclude the project in a way that reflects well upon the team leader, team members, and the organization performing the project. It is the time for

1. Determining that the project is, in fact, complete
2. Conducting final project evaluation
3. Assessing final risk and opportunity
4. Performing administrative closeout
5. Generating lessons learned
6. Cementing client relations

It is also the time to

7. Reward team member performance
8. Celebrate project success

Chapter 12 and Chapter 14 describe the closeout activities.

Politics

All projects are political. Not understanding this can be a source of considerable pain to project managers. Politics is the art of power and influence. The performance of a project conveys status and prestige. The outcome of a project may dramatically or subtly shift the power base within an organization. All individuals are concerned with their personal power and influence. The sooner you realize this and act in ways that enhance this power and influence for the important people in your organization, the better.

People Skills versus Hard Skills

The ability to establish the project schedule, the budget, and control metrics, and to forecast final completion date and cost are among the many hard skills required of the project manager and the team. These are important to project success. However, the ability to establish relationships with customers and stakeholders plus the ability to encourage creativity, commitment, and cohesion among team members are also very important. Of the two, people skills are the more important. The project manager may delegate the work to estimate the project's completion date or calculate cost and schedule variance. However, leadership is not something that may be delegated.

Managing Project Information

Projects produce a lot of information. Each phase of the project produces information. Some of the information must be filed in a place and format that make it readily available during those infrequent times when it is needed. Some information must be close at hand at all times. Some information must be disseminated or else it loses its value to the project. Knowing what information will be or should be generated at each phase of the project is important to information management. How information must be stored, retained, or disseminated is also important. Chapter 8 discusses managing project information.

Chapter 2

The Initiation Phase

Mega Recipe for the Initiation Phase

It would be best if the project manager is involved in the following five activities that should be achieved in the Initiation Phase, but unfortunately this is often not the case.

Activity 1: Determine current and desired conditions. Explore among sponsors, customers, and constituencies the need for the project. Develop comprehensive statements of: (1) why the project is necessary — the problems or unfulfilled needs that constitute the current condition, and (2) the desired conditions relative to each problem or unfulfilled need.

Activity 2: Write the business case definition. This outlines the potential project in all its facets so the team and its sponsors understand the full complexity of the project.

Activity 3: Estimate the potential project's economic viability — the return it will bring on the investment.

Activity 4: Brief the business case definition including its estimated economic viability to the sponsors (and constituencies if appropriate). Get clarifications and decisions on issues in the project definition that need to be finalized. Get a go or no go decision.

Purposes of the Initiation Phase

The Initiation Phase is the first of the four phases in the project life cycle. Its purposes are to assure that the needs for the project, its outputs, and the conditions under which it will be performed are clearly understood by the project's sponsors, and the project team and the sponsors have the same vision of the project.

Role of the Project Manager

The project manager's role is to lead the startup team through the four activities listed earlier. The project manager's involvement in these early activities helps to educate the manager and serves as the basis for building important rapport with the project's sponsors and major constituencies. Rapport building in this early phase of the project will pay dividends when the project manager needs help (technical, financial, or political) from one or more of the sponsors or constituency leaders. In addition to performing the four activities, the project manager seeks to sense the real desires of each sponsor and their level of commitment to the project.

Outputs of the Initiation Phase

1. Current needs and problems impacting the organization
2. The desired conditions that the project is to produce
3. The clarified business case definition
4. The decision by upper management and the sponsors to proceed or not proceed with the project

Someone in upper management, often the project sponsor, issues the charter. The charter is discussed later in the chapter.

Initiation Phase Problems

1. The Initiation Phase is not performed, or is performed in a very truncated fashion. The organization issues a charter, which assigns a project manager, outlines the purpose of the project, and authorizes the project manager to start the planning process. The planning is started with neither a comprehensive statement of the need for the project nor a complete statement of the desired condition relative to each need or problem that the project is intended to solve. The sponsors' understanding of the current problems or unfulfilled needs necessitating the project and the desired solution to each is unknown.

   Among the first activities, the project manager should conduct a comprehensive effort to determine the needs or problems associated with the current undesirable condition. The desired condition relative to each currently unfulfilled need or problem must also be determined. Stating the desired condition is important because there may be and often is more than one condition that may satisfy a need or problem. The project objectives list those desired conditions that the project is intended to produce. Complete the business case definition and brief these to the project's sponsors and get their clarification and approval.

2. The decision to launch the project did not include an assessment of economic viability. So the organization invests in a project with less payback than possible from an alternative. When upper management recognizes this, it kills the project. This in turn reduces employee morale and productivity.

   The project manager should have an economic viability analysis of some kind performed. Brief this to upper management before planning goes too far. As a minimum, the project manager should determine the project budget (cost) early and get this approved. Any "pushback" (non acceptance of the budget) from the sponsors or upper management does not augur well for the project's survival.

3. The project manager is assigned to run an effort justified via the "hunch technique." The hunch technique is where some high-ranking person (Mr. or Ms. Big) has an intuitive feeling that the project is needed but no needs assessment or definition has been performed.

   This problem requires a certain degree of political finesse. The project manager should perform the needs and desired conditions assessment plus the business case definition and privately brief Mr. or Ms. Big. The private briefing must not embarrass the "Bigs." This strategy has a certain risk attached to it. The "Bigs" may interpret your action as a challenge to their authority and vision. You need to indicate that the needs assessment and business case definition were performed "to get the team up to speed on the project — to fill in the gaps in our understanding." If in the process of discussing the project definition, etc., with the "Bigs," decisions are made that make the project more clear and achievable — so much the better! Be sure to thank the "Bigs" for their guidance and help.

4. During the briefing of the business case definition, one or more functional managers indicate that they cannot provide the people or other resources the project will require.

   This is a major threat to the potential project. Three choices are available: kill the project now, plan the project and wait until the resources are available, or plan the project based on resource constraints. Get a commitment as to the level of resources that can be provided and plan the project within these constraints. This will stretch out the project over a longer time. Plan out the project under the resource-constrained conditions and get it approved by all members of upper management. There may be one member of upper management who cannot or will not accept the protracted project schedule, and would force a higher level of resource commitment. (Remember your position here. Present the information and let the managers argue it out! Don't argue with someone who outranks you; it is not career enhancing to do so!)[1]

Activity 1: Determine Current and Desired Condition

All projects are launched because the organization believes there are compelling unfulfilled needs or important problems that must be solved. The unfulfilled needs or problems are part of the current condition. The solution to the unfulfilled needs or existing problems is part of the desired condition. Examples of needs include the need to improve production efficiency (more units per day or lower cost per unit) and the need to launch a new product or service to complete the organization's product or service line or to meet a competitor's challenge in the marketplace. Inadequate space in which to conduct operations is a current-condition problem. Having a factory space of 120,000 square feet is the desired condition — that is, a requirement of the solution. The project might achieve the requirement by expanding the current facility (building additional plant space) or by relocating to another facility. Before the project team can work out a solution, it must first understand the current problem and the desired condition (requirement).

Developing a specific, clear, comprehensive, commonly understood, and mutually agreed-upon statement of the needs or problems is the first step in defining a project. The statement must be clearly understood by all who read it. It cannot be written in technical language or jargon. Most of the time, the statement of need is written for upper managers and sponsors. It must be comprehensive so it articulates all the facets of the needs. Any part of the needs statement that is vague or left unstated leaves room for misinterpretation or political maneuvering. All the project sponsors must agree that the needs/problems statement is complete and accurate. Clarifying the needs statement requires considerable effort and time. It is better to pay this price during the initiation phase than it is to try to clarify the purpose of the project after the organization has invested in planning and execution. It is better to find out that the needs do not warrant a project effort during the initiation phase than it is to find out later when the price of discontinuing the project is much higher.

Likewise, an undefined or poorly defined desired condition will cause difficulty. There is usually more than one solution (desired condition requirement) to each unfulfilled need or current problem. Unless the team has clear information about the desired solution, it may develop a solution that is not acceptable to the project sponsors.

> **Example 1:** A factory in Newark, New Jersey, must improve its manufacturing capacity. This statement is clear but not very specific and not comprehensive. Does it mean that we need to conduct a study to identify those opportunities to improve capacity and efficiency? Does it mean that we need to replace the automatic "screw machines" because the existing machines can no longer produce parts to the necessary tolerances? Does it mean that we need to install more machines simply

to increase capacity or do we also need to increase the flexibility to produce a wider variety of items? Does the assembly line need to be upgraded also? Simply increasing the capacity to produce more component parts will not in itself increase the facility's capability and efficiency in assembling components into finished products. Before we can proceed to define the project further, we must determine the current conditions describing the needs/problems that must be addressed and the desired condition requirements of the solution. Once these are properly defined, we need to get the sponsors to approve.

How do you develop statements of current and desired condition? Where does the information come from? The information may come from records or files but mostly it will come from interviews. In the case of example 1, whom should the project team interview to determine the needs or problems that the project must address? The plant superintendent can provide information of the production problems plus any plans for changes to the physical plant. The chief engineer can provide information about the specific kinds of parts and components for which capacity must be increased and by how much as well as the dimensional tolerances that must be maintained. The chief production engineer can provide information about the current capacity and the various ways that capacity and flexibility need to be and can be increased. This person will be very familiar with the various approaches to increasing capacity, i.e., the kinds of technology and processes needed versus the technology and processes available in the marketplace. The machine shop foreman can provide information on the specifics of the production problems; this is the person who lives with the production problems on a daily basis. The warehouse supervisor can provide information on the impact increased capacity will have on the facility's ability to store these additional parts and finished items. Perhaps the warehouse can store additional component parts but it has limited space to store finished products, which would require a heavy-duty shelving structure that does not now exist. Perhaps increasing the capacity to store finished products will require new or additional product-moving equipment or perhaps a currently nonexistent automated inventory system. Maybe the problem is inventory control rather than manufacturing capacity. The head of sales can provide information on marketing plans to sell finished products that are not now manufactured or for which the warehouse has no storage capacity. The chief of finance can provide information on whether the organization's cash flow can finance the project or how big a change it can finance. All of these people can provide information about the desired condition relative to capacity, flexibility, storage space requirements, etc. This information is important to determining the need for change, the size of the change, the specifics of the desired change, etc. The information gathered during the interviews will be incorporated in the needs/statement of the current condition and the requirements part of the desired condition.

Individuals who are interviewed will usually feel included and part of the effort. Individuals who are not interviewed often feel left out of the information loop and unimportant; these people may and often do disparage or obstruct the effort because of this perceived alienation. It is usually better to include some perhaps unnecessary interviews than it is to omit a politically important person from the interview list. Interviewing is not just an information-gathering technique; it also serves to build a community of involved individuals who each possess, in varying degrees, a sense of ownership of the project.

Depending upon the kind of project, reviewing records may be a good source of information about the need for a project. Reviewing production records, production problems, after-action reports, sales reports, and accident reports are some potential sources. Reviewing records will not produce the favorable ego involvement results that interviewing often does.

An alternative approach for getting the current condition, desired condition, and project objectives is for the project manager to facilitate a small group discussion among those who understand the current problems and unfulfilled needs. (Small-group techniques are discussed in activity 2 of Chapter 9.)

Regardless of the technique used, the strategy is as follows:

Step 1: Discuss and get a comprehensive list of unfulfilled needs or problems with statements of the consequences or impacts of each.

Step 2: Discuss the desired condition relative to each need or problem articulated in step 1. Desired condition may be expressed in terms of cost, quality, appearance, features, capabilities, or functional attributes.

Step 3: Discuss and list the desired conditions from Step 2 that must be achieved in the proposed project. These constitute the project objectives. See Example 1 "Current Condition" and "Desired Condition" in Activity 2.

Activity 2: Write the Business Case Definition[2]

The purpose of the project business case definition (sometimes referred to as the project definition) is to provide upper management with all the information it needs to make an informed decision to pursue or forgo the project. Organizations invest in projects for purposes of generating some desired condition or result, or eliminating some undesired condition. The kind of information required to make these decisions varies from organization to organization and from industry to industry. The one thing all business case definitions have in common is that they seek to provide management with the information it needs to make the decision to invest or not invest in the project. The outline for the business case definition is shown in Table 2.1. The following topics are suggested; others may be added.

Table 2.1   Business Case Definition Outline

The purposes of the business case definition are to provide upper management the information it needs to make an informed decision about funding the project, and to assure that the project team and upper management have the same understanding of the project — in all its facets.

Current condition	Develop a comprehensive and detailed statement of the current condition, listing the problems and unfulfilled needs including the impacts and consequences of the problems and needs as perceived by the stakeholders. Problems and needs may focus upon cost, quality, appearances, features, capabilities, or functions that are needed but the organization lacks or that need to be changed or fixed.
Desired condition	Describe the cost, quality, appearances, features, capabilities, and functions that are desired. Be sure to include a desired condition for each problem or need listed in the current condition statement above.
Project objectives	List and describe the desired conditions from above that the proposed project is to achieve. Be sure to include the extent to which the desired condition will be achieved. Indicate if any desired conditions cannot be fully achieved in this project. Indicate any desired conditions from above that are specifically excluded from the proposed project objectives.
Economic viability	Estimate the cost and return of the proposed project using any of the following metrics: benefit to cost index, profit, cost savings, EVA, etc. (See activity 3 in Chapter 2.)
Deliverables	List all the deliverables that the project must produce. Deliverables may be tangible or intangible. Each deliverable must be described in sufficient detail to allow the reader to assess the challenge associated with producing each deliverable. ("Deliverables" refers to products or services that are produced and submitted to the customer.)
Customer-driven milestones	List the customer-determined significant events and their milestone dates.
Organizational expertise and capacity	Describe how the organization does or does not have the technical expertise, available people, and production capacity to perform the project. Indicate whether the project will require the organization to develop new expertise, or hire workers or subcontractors.

(Continued)

Table 2.1   Business Case Definition Outline (Continued)

Assumptions	An assumption is a condition that will impact both planning and execution of the project. An assumption once stated must be investigated to determine whether or not it is valid. Assumptions affect the scope, time, and cost of the project.
Constraints	A constraint is a condition or requirement that will limit the team's flexibility to perform the project. Constraints must be investigated to determine whether or not they are valid because they affect scope, time, and cost of the project. Constraints may include national and state regulations.
Potential problems	A potential problem is anything that may occur that has a negative consequence for the project: increased cost, safety hazard, delayed schedule, physical obstacle to work, etc. List all identified potential problems to performing the project.
Potential opportunities	An opportunity is anything good that might occur to decrease cost, increase revenue, circumvent obstacles, accelerate the schedule, etc.
Key stakeholders and constituencies	A stakeholder is anyone with a vested interest in the project and the power to help or hinder the project. A constituency is an organized group of people with a vested interest in the project. Sometimes stakeholders and constituencies want a project to succeed; sometimes they work to see it fail. List the stakeholders and constituencies plus the issues that are important to each.
Unusual resource requirements	Resource requirements include people with specialized training or experience, plus materials or equipment not available in the organization. These unusual skill sets, materials, or equipment will require a special process to obtain for the project.
Project performance criteria	List the criteria by which the customer or sponsor will judge the success of the project. The project team will write a tentative list of performance criteria and verify these as soon as possible with the customer or sponsor.

Current Condition

This paragraph describes the needs or problems that the project is intended to solve. The needs or problems must be described in sufficient detail so upper managers understand the extent of the unfulfilled needs or problems and their consequences. The needs statement must be comprehensive so the breadth of the undesirable current condition is apparent.

For Example 1, the current conditions are as follows:

1. Small part component production capacity is not adequate to meet the needs of the assembly line.
2. Warehouse is often not able to keep track of the many component parts required to supply our assembly line.
3. Assembly line is geared to putting together small products (toaster size); we need the capacity to assemble washing machine-size units.
4. Automatic screw machines have the capacity to produce 10,000 small "turned" parts per day, which is inadequate because so many are rejects; they can no longer produce parts to dimensional tolerances of plus or minus five thousands (±.005) of an inch.
5. We lack the capacity to "box and crate" and store washing machine-size units.
6. Warehouse can only move and store 300 large units per day.

Desired Condition

This paragraph describes the condition of acceptability — the requirements of capability, capacity, cost, quality, appearance, etc., that are desired relative to each need or problem in the statement of current condition.

For example 1, the desired conditions are as follows:

1. The assembly line can assemble 500 small units per day, where each unit requires no more than 30 operations to assemble.
2. Warehouse has an automated inventory and location system, which gives it the ability to know in "real-time" the quantity of components in inventory, those allocated, and those in the production schedule.
3. A second assembly line exists with the ability to assemble 400 washing machine-size units per day where each unit requires 40 operations or less to assemble.
4. Automatic screw machines can produce 10,000 units per day with tolerances held within ±.005.
5. We have the facility to box and crate 500 washing machine-size units per day.
6. Warehouse has the capacity to move 500 large units per day and store 10,000 large units.

Project Objective

This paragraph lists the desired condition (from the previous paragraph) that it will seek to achieve. The desired condition may include requirements that a single project may not be able to achieve. Perhaps the desired condition can be achieved by a program of projects funded over a period of time. In this case, the part of the problem that the proposed project will solve must be specified.

For example 1, the objectives of the proposed project are as follows (as previously described in "Desired Condition"):

2. Achieve a warehouse that has the automated inventory and location system that gives it the ability to know in "real-time" the quantity of component parts in the inventory, those allocated, and those in the production schedule. The capability to enter data and read data will exist in both the warehouse office and the production control office.

5. Build a facility to box and crate 500 washing machine-size units per day. This facility will be housed in the warehouse addition area.

6. Expand the current warehouse by 30,000 square feet. Purchase the forklifts or other suitable product-moving equipment so it has the capacity to move 500 large (washing machine size) units per day and store 10,000 large units; this desired capacity is in addition to that already existing.

The other desired conditions will be addressed by a subsequent project.

Economic Viability

This section must make the case for the economic advantage that the project will produce. It must describe how the cost of the project is worth what it will produce. The techniques for demonstrating this include the following:

1. Benefit to cost index
2. Simple profit
3. Multiple-year profit
4. Cost savings
5. Internal rate of return
6. Economic value added

These techniques are described later in this chapter.

Deliverables

This section lists and describes all the proposed project deliverables. Its intent is to inform upper management of the full range of products and services that the project must produce. Sometimes products or services associated with the project are outside the organization's core competency. For instance, a project to produce a computer software system may be within an organization's capability but the requirements to deliver a user-friendly manual, to produce training materials, and to conduct training may require outside expertise. The purpose of the business case definition is to assure that management is aware of these issues before it makes the decision to proceed with the project.

Customer-Determined Milestones

A milestone is an important event with a date connected to it. Under this paragraph, list all milestones imposed by the customer or sponsor.

Organizational Expertise and Capacity

Indicate here whether the organization has the in-house expertise to perform the project or if it must hire outside talent. Indicate also if the organization has the capacity — number of workers or production capacity — to perform the project.

Assumptions

An assumption is a circumstance or condition that impacts the "do-ability" and cost of the project. List all assumptions so that upper management may be informed. All assumptions, once stated, must be validated because a false assumption may constitute a problem for the execution of the project.

> **Example 2:** For a project to install computer software at all employee workstations, we assume that the installation may be performed during regular work hours. We are making this assumption because installing during non-regular work hours would require us to pay installers a differential for working during non-regular work hours (e.g., the night shift or weekends). Upper management needs to be informed of this assumption and the consequences (additional installation cost) if we are not allowed to install during regular work hours. All assumptions must be stated; once stated, all assumptions must be validated. That is, the correctness must be determined. Assumptions must be validated because they have implications for cost, time, money, and other resources.

Constraints

A constraint is a condition or circumstance that limits the project team's flexibility to perform — perhaps increases the project time and cost.

> **Example 3:** The solicitation for a project to build a town park includes the requirement that all workers must be members of the local laborers and masons unions. This requirement is a constraint that reduces the contractor's flexibility to hire workers. It also establishes the wage scale for both these categories of workers. Whom the contractor may hire

and the wages to be paid are constraints placed upon the contractor. Sometimes it is possible to get around a constraint; however, doing so costs money, and the "get-around action" must be ethical and legal.

Example 4: The weather imposes constraints on the time of year when outdoor construction may be performed. Outdoor construction, except emergency construction or repair, is usually curtailed during the cold-weather months because of reduced human performance and increased safety concerns. Naturally, upper management needs to know of any constraints associated with the proposed project. The cost of providing shelter from the elements must be added to the cost of the project if the schedule requires outdoor work to be performed during the cold weather.

Potential Problems

During the initiation phase, we do not have a definitive risk management plan that identifies project risks. (The writing of the risk management plan occurs during the planning phase.)

A potential problem is anything that may occur that has negative consequences for the project: increased cost, delayed delivery, reduced quality or customer satisfaction, etc. Describe all potential problems associated with the project so upper management may be informed of these before it decides to approve the project.

Potential Opportunities

An opportunity is anything that may occur that would have a positive effect upon the project: decreased cost, shortened schedule, increased revenue, increased goodwill, etc. List all the potential opportunities so upper management may consider them in addition to the potential problems.

Key Stakeholders and Constituencies

A stakeholder is a person who has a keen vested interest in the outcome of the project. This interest may seek the project's success. Some stakeholders will want to see the project fail. A constituency is a group of people with a commonly shared interest in the project. Sometimes constituencies have considerable power to place constraints upon the project or to jeopardize the project. Other constituencies are supportive of the project.

Example 5: There are many constituencies to the building of a nuclear electricity-generating plant. Environmental groups will be concerned with the impact the plant will have on local wildlife, air, and water quality. Industry groups will want the project because it will produce the electrical power needed to sustain and expand the manufacturing base. Remember that all projects are political. Identifying all the constituencies to the project and including their needs and priorities into the business case analysis is important to the decision to proceed with the project. Sometimes the political threats or fallout from a project make it untenable.

Indicating the stakeholders and constituencies to the project provides upper management with the information with which to judge the political feasibility of the project. A project has little chance of success if the majority of political interest and power is opposed to the project.

Unusual Resource Requirements

Upper management needs to be informed when a proposed project requires skills and expertise that do not exist within the organization or are in scarce supply. These unusual resource requirements will have to be hired or contracted, which is appropriate as long as the external expertise is available at a reasonable cost. Hiring outside resources has advantages and disadvantages. Upper management needs to know the extent to which the proposed project depends upon external or unusual resources.

Project Performance Criteria

The customer will evaluate the project team's performance along many dimensions. Listing these criteria will help upper management understand the full weight of the customer's expectations. Knowing the customer's acceptance criteria (these may be called ancillary requirements) will help the team succeed.

> **Example 6:** An organization receives a solicitation for a project to renovate a large tourist hotel. The solicitation includes specifications for the renovation, plus a list of other criteria by which the contractor's bid will be evaluated. These evaluation criteria are really ancillary requirements placed upon the contractor. The customer wants to keep the hotel operational during the renovation so there are many ancillary requirements in addition to the obvious requirements of scope, cost, time, and quality of workmanship. Identifying these evaluation criteria is important because they impact the cost and time to perform the project as well as risks. Knowing the project performance criteria is necessary to the decision to pursue the contract.

If it does not already exist, the project team will create a list of draft criteria, and get the actual criteria from the customer as soon as possible. Possible questions: What measures will you use to judge the success of this project? What criteria are important to you in assessing project performance? Frequently, these questions will generate ideas about what is important to the client but were completely unknown to the team.

Charter

Organizational members must be informed when a project has been approved. Someone in upper management, often the project sponsor, issues the notification. "Charter" is the name given to this document. The format and information in the charter vary from organization to organization but typically they include the following:

1. Official name of the project
2. Brief statement of the purpose of the project
3. Name of the project manager and names of supporting team members
4. Role and responsibility of the project manager
5. The authority of the project manager
6. List of documents and deliverables required of the project

Sometimes it will include

7. A tentative schedule
8. A budget

In addition, the charter will sometimes include a diplomatic appeal for functional managers to support the project by indicating the project's importance. The approved and clarified business case definition should be part of the charter because it documents what the project must produce and the circumstances and conditions incident to the effort.

Activity 3: Estimate the Project's Economic Viability

This section describes techniques for estimating the economic payback from a potential project.

Benefit to Cost Index

The benefit to cost approach (BC) requires that the analyst estimate the benefit and the cost of the project. The benefit to cost index is simply the result of dividing the benefit by the cost. Benefit and cost may be expressed in dollars or any other units.

$$\text{B/C index} = \text{benefit} \div \text{cost} \tag{2.1}$$

Example 7: The project is estimated to generate a benefit (e.g., revenue) of $100,000. The cost is estimated at $50,000. The benefit to cost index is 2 ($100,000 ÷ $50,000). This is interpreted to mean that for every dollar invested, the organization expects a benefit of two dollars.

The benefit to cost indices may be used to rank the economic viability of three similarly sized projects.

Example 8: Proposed projects A, B, and C have benefit to cost indices of 2, 3, and .75, respectively. The costs and benefits of each are shown as follows:

Project	Benefit ($)	Cost ($)	B/C Index	Rank
A	100,000	50,000	2.00	2
B	150,000	50,000	3.00	1
C	37,500	50,000	.75	3

Project B has the highest benefit to cost index because it is estimated to produce a benefit of three dollars for every dollar of cost; it is ranked first. Project A and project C are ranked second and third because they are estimated to produce a benefit of two dollars and seventy-five cents, respectively, for each dollar of cost. Benefits and costs do not have to be expressed in dollars.

Example 9: Clinic A is estimated to have a capacity (benefit) for treating 3,000 patients per year at a cost of $300,000 per year (including labor, medical supplies, and medicine). Clinic B is estimated to have a capacity of 5,000 patients per year at a cost of $800,000 per year. Both clinics treat patients who are not able to pay for their treatment.

The benefit to cost index for clinic A is .01 (3,000 patients served ÷ $300,000). This means that each dollar of cost will "pay for" .01 patients treated. The index can be expressed as .01 ÷ $1. (Multiplying the top and bottom of a fraction by the same number does not change the value of the fraction.) Multiplying the top and bottom by 100 produces the fraction 1/100; that is, B divided by C equals 1 patient per $100. This means that every patient served in clinic A costs $100.

The benefit to cost index for clinic B is .00625 (5,000 patients served ÷ $800,000). This means that each dollar of cost will "pay for" .00625 patients treated. The index can be expressed as .00625 ÷ $1. Multiplying the top and bottom by 160 produces the fraction 1/160. This means that every patient served in clinic B costs $160.

Comparing the two indices indicates that clinic A has a stronger economic viability than does clinic B because the cost per patient served in clinic A is $100 versus $160 in clinic B. (Of course, we have already determined that the types of illnesses treated in each clinic are similar.)

Simple Profit

Simple profit is also called single-year profit. Simple profit is defined as revenue minus cost.

$$\text{Simple profit} = \text{revenue} - \text{cost} \tag{2.2}$$

This technique is a common way to estimate the economic viability of a project that can be completed in one year or less. Revenue and cost are expressed in units of currency (U.S. dollars for projects in the United States). This technique is not sensitive to the time value of money, so it should not be used for multiple year projects. See the section "Present Value of Money."

> **Example 10:** An organization has two opportunities to generate profits. Project A has an estimated revenue of $5 million with a cost of $4 million. Project B is expected to bring a return (revenue) of $6.5 million at a cost of $4 million. Both projects will take less than a year. The profit for project A is estimated at $1 million ($5 million – $4 million). Project B has an estimated profit of $2.5 million ($6.5 million – $4 million). Of the two, project B has the stronger economic viability. It generates more dollars of profit than does project A even though the investments (costs) are equal.

Present Value of Money

Present value (PV) of money is not a technique for assessing economic viability; it is a technique for converting dollar amounts in various out-years to a common base called the present (i.e., today). The need to convert dollar amounts to present value is based on the idea that a dollar today is worth more than a dollar one, two, or more years from now. If a person has the choice of getting $1,000 today or waiting five years, the obvious choice is to take the money today. The reason: $1,000 invested in a bank will be worth more than $1,000 at the end of five years. The idea that money has different values depending upon when you get the money or when you spend the money is called the time value of money. The need to consider the time value of money applies to multiple-year projects.

The formula for converting dollar amounts to their present value is

$$PV = FV \div (1 + I)^y \tag{2.3}$$

where PV means present value, FV means future value, the letter I is the interest banks are paying, and *y* equals the number of years from now (the present) that the money will change hands. Dollar amounts that have been converted to their present value are referred to as normalized or discounted dollars.

> **Example 11:** A person tells you that he will pay $1,000 for your vintage camera but you must wait one year before he can give you the money. Ignoring the risk that he may default on paying you a year from now, what is the present value of the $1,000 that you will get a year from now? Interest rate is 5%.
>
> $$PV = FV \div (1 + I)^y$$
> $$PV = 1000 \div (1.05)^1$$
> $$PV = \$952.38$$

Organizations must use normalized cost and revenue figures to calculate profitability of multiple-year projects. Doing so is called net present value of profit.

Multiple-Year Profit (Net Present Value of Profit)

Net present value (NPV) is the term for profitability of a multiple-year project. NPV equals the total present value of revenues minus the total present value of costs. All multiple-year projects should have their potential profit calculated via this NPV technique:

$$NPV = PV \text{ of revenues} - PV \text{ of costs} \tag{2.4}$$

> **Example 12:** An organization has estimated the costs and revenues for a five-year project. The effects of inflation and increased costs are reflected in the cost estimates. In the following chart, all figures are in dollars; the interest rate is 5%.

Year	Cost	PV of Cost	Revenue	PV of Revenue
1	50,000	47,619	0	0
2	55,000	49,887	80,000	72,562
3	60,000	51,830	90,000	77,745
4	66,000	54,298	120,000	98,724
5	70,000	54,847	150,000	117,529
Totals		$258,481		$366,560

The estimated NPV (profit) for this project is $108,079 ($366,560 − $258,481). Notice that it is not acceptable to use the un-normalized total cost and revenue figures to calculate profit.

Payback Period

Payback period answers the following question: How long must we wait to recoup our investment? This technique is usually used when there is a single up-front investment in the project. It seems less useful and less frequently used for a project having multiple years of investment. The formulas are

$$\text{Investment amount} \div \text{profit per month} = \text{payback period in months} \quad (2.5)$$

$$\text{Investment amount} \div \text{profit per year} = \text{payback period in years} \quad (2.6)$$

> **Example 13:** A petroleum refinery is estimated to cost $150 million. It will take two years to build. Once operating, the refinery will generate $10 million of profit per month. How long is the payback period? Stated another way, how long after the investment is complete (refinery is completed) will we have to wait to recoup our entire investment? Payback period equals 15 months ($150 million ÷ 10 million/month).

Return on Sales

Return on sales (ROS) is a technique for estimating economic viability where the organization will sell the output of the project. ROS is expressed as a fraction: ROS = return (i.e., net profit) ÷ sale price. Net profit is profit after taxes have been paid on the profit. Sale price equals cost + profit desired by the organization. The unit of measure for both sale price and net profit is dollars. This technique is not sensitive to the time value of money so it should not be used for multiple-year projects. (See the section "Present Value of Money.")

$$\text{ROS} = \text{net profit} \div \text{sale price} \quad (2.7)$$

$$\text{ROS} = (\text{profit} - \text{taxes}) \div \text{sale price} \quad (2.8)$$

> **Example 14:** An organization needs to choose between two projects. Project A involves producing a system that will sell for $1 million and is estimated to cost $760,000. The profit is estimated at $240,000 ($1,000,000 − $760,000). The organization pays 35% taxes so the net profit is $156,000 ($240,000) − (.35)($240.000). The return on sales is

.156 ($156,000 ÷ $1,000,000) or 15.60%. Project B involves producing another system that will sell for $1.5 million and is estimated to cost $1 million to produce. The before-tax profit is estimated at $500,000 ($1,500,000 – $1,000,000). The after-tax profit is $325,000 ($500,000) – (.35)($500,000). The return on sales is .2167 ($325,000 ÷ $1,500,000) or 21.67%.

Project B has the stronger economic viability because it is estimated to produce a 21.67% return on sales whereas project A is estimated to produce 15.60%.

Return on Investment

Return on investment (ROI) determines the percent of profit over the cost of the project.

$$\text{Return on investment (ROI)} = \text{before-tax profit} \div \text{cost} \qquad (2.9)$$

For project A in example 14, the ROI is .3158 ($240,000 ÷ $760,000) or 31.58%. For project B, the ROI is .50 ($500,000 ÷ $1,000,000) or 50%. ROI is a way of expressing percent profitability over cost. Project A is estimated to give a 31.57% return and project B is estimated to provide a 50% return on the investment. Stated another way, project A has an estimated profit of $.3157 for each dollar invested and project B has an estimated profit of $.50 for each dollar invested.

Net Cost Savings

Some projects are initiated because the organization needs to reduce costs; the purpose of the project is to generate cost savings.

$$\text{Net cost savings} = \text{total costs saved} - \text{total project cost} \qquad (2.10)$$

Example 15: A project is estimated to generate a cost savings of $3 million over the one-year life of the project. The project cost is estimated at $1 million. The net cost savings are estimated at $2 million ($3 million – $1 million).

The benefit to cost approach (discussed earlier) is an alternative method to estimate the economic viability of this one-year project. For a multiple-year project, the benefit to cost approach is not appropriate because it doesn't consider the time value of money.

Example 16: A three-year project is estimated to generate the costs and cost savings listed in the following chart. Interest rate on money is 5%. All dollar values must be converted to present value.

Year	Cost ($)	PV of Cost ($)	Cost Savings ($)	PV of Cost Savings ($)
1	50,000	47,619	0	0
2	150,000	136,054	400,000	362,812
3	75,000	64,788	800,000	691,070

Total PV of Cost = $248,461
Total Cost Savings PV = $1,053,882
Net Cost Savings = $1,053,882 − $248,461 = $805,421

This three-year project has an estimated present value cost of $248,461 and an estimated present value total cost savings of $1,053,882. This produces an estimated present value net cost savings of $805,421.

Internal Rate of Return

Internal rate of return (IRR) is used to assess the economic viability of multiple-year projects. It is the percentage of return expected from the organization's use of its internal money — hence the name "internal rate of return." It is defined as the percent number used in the present value formula where the total normalized costs of a project equal the total normalized revenues of the project. There is no direct way to determine the percentage that is the internal rate of expected return. A process of successive approximations is used to determine the IRR.

Example 17: An organization estimates the costs and revenues of a five-year project as follows:

Year	Cost ($)	PV of Cost	Revenue ($)	PV of Revenue
1	50,000		0	
2	55,000		80,000	
3	60,000		90,000	
4	66,000		120,000	
5	70,000		150,000	
Totals				

Determining the IRR for this project means finding the I in Equation 2.3: $PV = FV \div (1 + I)^y$, where the total PV of costs equals the total PV of revenues. We have no way of knowing what interest rate will produce normalized costs equal to normalized revenues so we guess at a percent number. We see what this first guess produces and then take another guess. Each guess will get us closer to that percent number that produces total normalized costs that equal total normalized revenues.

The first guess is 5%. The PV formula $PV = FV \div (1 + I)^y$ provides the normalized figures shown in the PV columns.

Year	Cost	PV of Cost	Revenue	PV of Revenue
1	50,000	47,619	0	0
2	55,000	49,887	80,000	72,562
3	60,000	51,830	90,000	77,745
4	66,000	54,298	120,000	98,724
5	70,000	54,847	150,000	117,529
Totals		$258,481		$366,560

The normalized cost of $258,481 does not equal the normalized revenue of $366,560. The difference is $108,079. The next guess is 15%. The following table shows the results of using 15% to normalize the cost and revenue figures.

Year	Cost	PV of Cost	Revenue	PV of Revenue
1	50,000	43,478	0	0
2	55,000	41,488	80,000	60,491
3	60,000	39,451	90,000	59,176
4	66,000	37,736	120,000	68,610
5	70,000	34,802	150,000	74,577
Totals		$196,955		$262,854

The difference between the total normalized costs and normalized revenues is $65,899. We are still a long way away from an interest rate that produces equal normalized costs and revenues.

The process of guessing at the IRR percent number and determining the resultant difference can take a lot of time. Fortunately, we have an Excel© spreadsheet that accelerates the process. Figure 2.1 is an example of the spreadsheet-generated IRR analysis. (The template for automated calculations is shown in Figure 2.2.) The following table summarizes the odyssey to find the interest rate where the cumulative normalized costs equal the cumulative normalized revenues.

	A	B	C	D	E	F	G
Year		Cost	PV of Cost			Revenue	PV of Revenue
1		$50,000	$30,565			$0	$0
2		$55,000	$20,553			$80,000	$29,895
3		$60,000	$13,706			$90,000	$20,559
4		$66,000	$9,216			$120,000	$16,757
5		$70,000	$5,975			$150,000	$12,804
6			$0				$0
7			$0				$0
8			$0				$0
9			$0				$0
10			$0				$0

Total PV of Cost = $80,016   Total PV of Revenue = $80,016

Trial Interest Rate = 0.63586

Difference: PV of Revenues − PV of Costs = $0

Figure 2.1   Example of IRR analysis. The IRR has been determined when the difference between the PV of revenue and the PV of cost equals zero. Therefore, in this example the IRR = 63.586%.

Interest (%)	Normalized Costs ($)	Normalized Revenues ($)	Difference ($)
5	258,480	366,560	108,080
15	196,955	262,854	65,899
40	115,835	132,742	16,907
60	84,129	85,839	1710
61	82,945	84,155	1210
63	80,664	80,927	263
64	79,564	79,380	−184
63.6	80,000	79,994	−6
63.586	80,016	80,016	0

Notice that the differences are positive for all percentages from 5% to 63%. The difference for 64% is negative, indicating that 64% is too high. The template in Figure 2.2 (included in the compact disk) helped us to find the exact percent number quickly. The IRR for this project is 63.586%.

A	B	C	D	E	F	G
Year	Cost	PV of Cost			Revenue	PV of Revenue
1		$0				$0
2		$0				$0
3		$0				$0
4		$0				$0
5		$0				$0
6		$0				$0
7		$0				$0
8		$0				$0
9		$0				$0
10		$0				$0

===

Total PV of Cost = $0   Total PV of Revenue = $0

Trial Interest Rate =   [   ]

Difference: PV of Revenues - PV of Costs =   [$0]

===

Figure 2.2   Spreadsheet template for IRR analysis.

Instructions:

(1) The template is designed to handle up to 10 years of data. If necessary, add rows for the additional years. If project is 10 years or less, skip steps 2 through 6.

(2) In column A, add the numbers for the additional years. For example, years 11, 12, 13, 14, and 15 would be in rows 17, 18, 19, 20, and 21, respectively.

(3) Copy the formula from column C downward so the formula applies to the added rows. For example, in column C change the formula for year 11 to = C17/(1+D23)^11. (The ^ symbol is the symbol for exponent. Thus, = C17/(1+ D23)^11 means the number in cell C17 is divided by the number 1 plus the number in cell D23 where the sum of 1+D23 is raised to the 11th power.) For year 12, C17 will become C18 and ^11 will become ^12. D23 is the same for all years. Do this for all the added years.

(4) Copy the formula from column G downward so the formula applies to the added years. For example, in column F change the formula for year 11 to = F17/(1+D23)^11. For year 12, F17 will become F18 and ^11 will become ^12. D23 is the same for all years. Do this for all the added years.

(5) If you have added years, change the formula in cell C18 from = SUM(C6:C16) to = SUM(C6:C21).

(6) If you have added years, change the formula in cell G18 from = SUM(G6:G16) to = SUM(G6:G21).

(7) Indicate the estimated cost for each of the years in column B.

(8) Indicate the estimated revenue for each of the years in column F.

(9) Enter the trial interest rate in cell D23. Show the number as a decimal, e.g., .75, not 75%. The spreadsheet will determine the difference of PV of revenues minus the PV of costs in cell F25 after you hit the enter key.

Note: The internal rate of return (IRR) has been determined when the difference between the PV of revenue and the PV of cost equals zero. See the example in Figure 2.1.

Example 18: An organization has three multiple-year projects to consider. The IRRs for the projects are as follows. project A: 63.586%; project B: 45.045%; and project C: 34.889%. Project A has the greatest economic viability because the organization can expect a 63.586% return on its investment.

Economic Value Added

The economic value added (EVA) technique is similar to the simple profit approach. The name "economic value added" derives from the idea that a project that produces a profit is adding to the economic value or worth of the organization. EVA is another term for profit. The EVA technique determines the profit that a project is expected to generate but it does it in a slightly different way — by including the cost of raising the money (called capital) to fund the project as an additional expense over normal project costs. The cost of raising capital is called the cost of capital.

Organizations need money (capital) to fund projects. There are three sources of money: profits from organizational operations, borrowing money, and selling equity (part ownership) in the organization. There is a cost associated with getting money from each of these sources. Money taken from a bank account costs the organization the interest that it does not earn on the money withdrawn. Money borrowed from a bank or other lender costs the interest that the lender charges. Issuing preferred stock or selling organizational bonds are ways of borrowing money. Money raised by selling preferred stocks or bonds comes with a cost; this is the percent return the stock or bond owners expect.

Example 19: An organization needs $1 million to fund a project. It is estimated that the project will generate a profit before taxes of $3 million. The organization's tax rate is 40%. It raises the money by borrowing $400,000 from a bank at 12%; by withdrawing $200,000 from a bank account that pays 3%; and $400,000 by selling bonds where the bondholders expect a return of 6%. The cost of borrowing the $400,000 is $48,000 (.12 × $400,000). The cost of the $200,000 withdrawn from the bank is $6,000 (.03 × $200,000). The cost of the $400,000 raised by selling bonds is $24,000 (i.e., .06 × $400,000).

The capital required to fund the project is $1 million. The cost of raising the $1 million is $78,000 (i.e., $48,000 + $6,000 + $24,000).

$$EVA = \text{net profit after taxes} - \text{cost of capital} \qquad (2.11)$$

For this project, the EVA is

$$EVA = profit - taxes - cost\ of\ capital$$

$$Simple\ profit = revenues - cost$$

$$Simple\ profit = \$3,000,000 - \$1,000,000 = \$2,000,000$$

$$Net\ profit = simple\ profit - taxes$$

$$Net\ profit = \$2,000,000 - .40 \times \$2,000,000 = \$1,200,000$$

$$Cost\ of\ capital = \$78,000$$

$$EVA = net\ profit - cost\ of\ capital$$

$$EVA = \$1,200,000 - \$78,000 = \$1,122,000$$

The profit (called EVA) from this project will increase the value or worth of the organization by $1,122,000. It is possible for a project to produce an economic loss; this is called economic value loss (EVL).

An alternative approach may be used when an organization has many sources from which to raise money. Rather than calculate the actual cost of raising the money, it may calculate the weighted average cost of raising the money. If the organization is estimating the EVA for many projects, it is easier to apply the weighted average cost of raising capital rather than determining the actual cost.

Example 20: An organization traditionally funds its projects from a Projects account funded by three sources of money. It calculates the weighted average cost of capital (WACC) as follows:

Source	Amount Raised ($)	Cost (%)	Cost Amount ($)
Bond sales	1,000,000	6	6,000
Bank 1	3,000,000	12	360,000
Bank 2	2,000,000	15	300,000
Totals	6,000,000		666,000

WACC = $666,000 ÷ $6,000,000 = .111 or 11.1%. Anytime the organization funds a project from the Projects account, it uses 11.1% as the weighted average cost of capital.

Example 21: The organization estimates the cost and revenue of a project at $2,500,000 and $5,000,000, respectively. The tax rate is 40%.

The project will be funded by the Projects account, which has an 11.1% WACC. The profit expected is $2,500,000 ($5,000,000 – $2,500,000). The expected net profit after taxes is $1,500,000 (i.e., $2,500,000 – .4 × $2,500,000).

From Equation 2.11:

$$EVA = \text{net profit after taxes} - \text{cost of capital}$$

$$\text{Cost of capital} = \text{project cost} \times WACC = \$2,500,000 \times .111 = \$277,500$$

$$EVA = \$1,500,000 - \$277,500 = \$1,222,500$$

Subjective Techniques

Two subjective approaches to selecting a project for funding are described next. Unlike the numerical approaches, these techniques do not determine the economic viability of the project. They simply provide a means of ranking potential projects on one or more subjective scales of worthiness. The disadvantage of the subjective techniques is that, although they will indicate the best of the candidate projects, the best may not be economically viable. Economic viability requires estimates of cost and revenues or cost and benefits, etc. Launching a project requires a considerable and sustained commitment to fund the project. Without some measure of economic viability, the organization does not know what financial commitment it is making and what return it can expect.

Subjective Rating Sheet

The rating sheet technique uses a predetermined list of rating criteria. The project selection committee uses the rating sheet to determine a total composite score for each candidate project. Projects are ranked according to their composite scores.

Each organization must design its own rating sheet that assesses those worthiness measures relevant to the organization and the industry. The design of the rating sheet is crucial because it determines the criteria by which project worthiness will be measured. The various worthiness measures may or may not be equally weighted; this is a decision that must be made by the rating sheet designers. The organization may establish a minimum composite score with the understanding that projects failing to reach this minimum score will be dropped from further consideration.

Example 22: An organization uses the rating sheet in Table 2.2 to evaluate the relative worthiness of four candidate projects. A Likert-type rating scale is used: 1 (strongly disagree), 2 (disagree), 3 (slightly disagree),

Table 2.2   Example of Composite Rating[a] Sheet Method

Criteria	Weight	Rating	Composite
1. Need for project is clear to me	.1	2	.2
2. Organization needs this project	.1	3	.3
3. Full organizational commitment is clear	.1	1	.1
4. Project definition is clear to me	.1	2	.2
5. Risks are minimal or do-able	.1	3	.3
6. Project uses our proven technology	.1	4	.4
7. We can achieve the project within reasonable costs	.1	3	.3
8. We can meet the project schedule	.1	7	.7
9. We can achieve the required quality	.1	7	.7
10. Our relationship with the customer or sponsor is strong	.1	4	.4
Total rating score			3.6

[a] Rating scale: 1 = strongly disagree; 2 = disagree; 3 = slightly disagree; 4 = can't decide; 5 = slightly agree; 6 = agree; 7 = strongly agree.

4 (can't decide), 5 (slightly agree), 6 (agree), and 7 (strongly agree). The rating sheet uses 10 equally weighted criteria. The committee may average the rating of all members or discuss each criterion until a consensus rating is achieved. The latter method will produce better results because it requires members to discuss, understand, and justify their ratings. The consensus ratings for project A are shown in Table 2.2.

The total rating scores for projects A, B, C, and D are 3.6, 5.1, 6.4, and 4.5, respectively. Project A and project D are clearly perceived as marginal in worthiness because their total scores are in the slightly-disagree-to-neutral range (3.6 = slightly disagree, and 4.5 = can't decide). There is no strong commitment for project B because the committee only slightly agrees (5.1 = slightly agree) that it is worthy. Project C is perceived to be worthy because the total rating is 6.4; this is between agree and strongly agree. The projects are rank ordered: C, B, D, and A with only project C recommended for funding.

Paired Comparison

Paired comparison is another subjective technique for ranking a list of candidate projects. Refer to Table 2.3 while reviewing the following:

Step 1: Each combination of projects two at a time is listed in the "Comparison Pair" column. Thus each project must be paired with one other project so they may be compared. All combinations of projects two at a time must be considered. The number of pairs equals [N(N – 1)] ÷ 2. With four projects, there are six pairs ([4(4 – 1)] ÷2) that must be compared.

Step 2: Define the criterion "worthiness" by which each competition will be judged. Worthiness is defined, for example, as a project that can successfully solve an important organizational need at a reasonable time and cost. The difficult part of this technique is defining worthiness because each organization must define it in terms relevant to the organization.

Step 3: The team discusses each pair and decides which of the two is the more worthy project.

Step 4: Count the number of times each project has "won" in the comparisons. Create a frequency column to indicate this information.

Step 5: Rank each project according to its frequency.

Example 23: An organization has four candidate projects to rank but sufficient funding to launch only two. The selection committee uses the paired comparison technique to rate the relative worthiness of each project. In Table 2.3, the four projects are identified as A, B, C, and D. Each possible pair of combinations is listed in the "Comparison Pair" column. The decision of which project of the pair is more worthy is recorded in the "More Worthy" column. The frequency with which each project "won" its competitions with other candidate projects is shown in the "Frequency" column. Rank position is assigned according to frequency; the largest frequency is assigned rank 1, the second highest is assigned rank 2, etc.

If there is a tie in the "Frequency" column, the decision as to which of the two is ranked higher is made based on the results shown in the "Comparison Pair" column. For instance, if the "Frequency" column showed a 3 for both projects

Table 2.3 Example of Paired Comparison Method of Ranking

Comparison Pair	More Worthy	Frequency	Rank
A versus B	B	A: 2	2
A versus C	A	B: 3	1
A versus D	A	C: 0	4
B versus C	B	D: 1	3
B versus D	B		
C versus D	D		

A and B, B would be ranked 1 because B was judged more worthy than A in the "Comparison Pair" column.

Hunch Technique

The "I have a hunch" technique is where someone with sufficient authority decides that a project is necessary to meet an important need. An example is when someone decides to launch a new product based on a "feel" for the marketplace. Sometimes launching a project from a subjective position may be very successful; this is the stuff of which entrepreneurs are made but doing so is very risky because no discipline has been applied to the decision. The "I just think we should do it" technique is frequently used. The results are often discontinued projects and wasted resources.

Activity 4: Brief the Project Definition and Get the Go or No-Go Decision

In activity 2, the team defined the project in all its various facets. This included information and sometimes the team's best guess as to the needs, purpose, deliverables, assumptions, constraints, etc. The purpose of activity 4 is to get the unclear, ambiguous information clarified by upper management and then to get the go or no-go decision. Briefing the members of upper management at one session is probably the most efficient way to do this. The members in attendance will hear the ideas of other members and may challenge them. Alternative opinions will be surfaced and discussed. This is good if it eventually leads to clarification and a decision. The project manager will facilitate the briefing and discussion. The project manager may need to ask for clarification or a decision when upper managers avoid them. Try not to accept a postponement of decisions or clarifications because these mean a delay in the planning process or planning for the project when information antecedent to the planning process is missing.

An alternative method of achieving clarification and decision making is to staff the project definition among upper managers. This process will take much longer than the decision briefing because responses will occur over a protracted period of time. Some upper managers will not respond at all until reminded. The second disadvantage is that differences of opinion will need to be received and then forwarded again to upper managers for their reactions or conclusions. This reciprocating clarification and decisions process can take a lot of time — probably weeks, maybe months. Most of the time, waiting weeks or months for a project to be clarified and decided is not desirable.

There are many reasons for a project being selected for funding: some reasons are rational, objective, and driven by economic viability; the justification for other projects may rely upon disciplined judgment. Finally, some projects are launched simply because they seem to fit a need or solve a problem (even though no evidence has been generated to demonstrate this expectation). Projects with outputs consistent and supportive of the organization's strategic plan fall into this category. The ease or difficulty with which activity 4 is accomplished depends upon organizational culture, the power and authority of the project sponsor, the distribution of costs and benefits from the project, and the perceived necessity of the project.

Notes

1. The last job I had with the U.S. Army was organizational development consultant in the Pentagon. I had this great job for about eight years. It gave me the opportunity to help general officers in bringing about organizational change. One of the maxims for staff officers is, "When the elephants romp, the pigmies get out of the tall grass." This is translated as, "when the generals are making decisions or discussing or disagreeing, don't get caught in the middle!" For issue number 4, this is good advice.
2. The business case definition is similar in content and purpose to the project requirements document (PRD) tool and slide 2-22 in *Managing Projects,* January 2006, published by ESI International, 901 North Glebe Road, Arlington, VA 22203. With permission.

Chapter 3

The Planning Phase

Mega Recipe for the Planning Phase

Activity 1: Assemble the full team. The skeleton team that existed for the initiation phase (if there was one) needs to be expanded so it includes members from the major relevant disciplines, e.g., engineering, manufacturing, sales, procurement, accounting, etc.

Activity 2: Review (or develop) and discuss the business case definition; this is important because a thorough understanding of the project objectives, assumptions, constraints, potential problems, etc., are crucial to the planning process. Develop a business case definition if one does not exist; get it approved.

Activity 3: Develop the work breakdown structure (WBS). Get it approved if necessary. The WBS is an organized list of all the work that must be performed in the project. Chapter 4 describes how to develop this.

Activity 4: Develop the budget; get it approved. The budget is the sum of the estimated costs of the work packages listed on the WBS. Don't forget to include risk in the budget estimate. See Chapter 5.

Activity 5: Develop the project schedule; get it approved. Chapter 6 shows how to develop the schedule. It's a good idea to include risk in the schedule.

Activity 6: Develop the resource plan. The resource plan lists all of the resources [people, time (start and end dates), budget, materials, special tools, and equipment] needed to accomplish each work package. It is not necessary to get it approved because it consists of information already approved in activities 1 through 5. Chapter 7 discusses the resource plan.

Activity 7: Develop the communications plan. The communications plan seeks to build goodwill and cooperation among the project's stakeholders and constituencies:
1. It lists all the stakeholders and constituencies to the project.
2. It indicates the information needed or desired by each.
3. It indicates how the information needs of each will be provided.
4. It may be expanded into a stakeholder and constituency management plan if it includes information about the political issues, power, and goals of each and includes strategies for shaping or motivating this energy into support for the project.

A communications plan is probably needed on all projects although a stakeholder and constituency management plan is needed only on high-profile and highly politicized projects.

Activity 8: Develop the risk management plan. This is the last of the planning documents to be developed because the process involves examining the previous planning documents to identify threats and opportunities to the project. It includes a list of threats and another of opportunities plus strategies for dealing with each. Chapter 9 describes how to develop the risk management plan.

Purpose of the Planning Phase

The purpose of the planning phase is to develop the documents needed to execute and control the project. A second purpose is to provide upper management or sponsors the opportunity to give advice and approve the plan. Sometimes it is neither expected nor appropriate for the planning documents to be reviewed and approved by upper managers or sponsors. This depends upon the nature of the project, the experience of the project manager, and the culture of the organization. Having an external customer review and approve the planning documents may not be appropriate because doing so establishes the approved documents as part of the baseline. However, customer approval of planning documents is required on large, expensive, and politically sensitive projects such as building an aircraft carrier. (See Chapter 10.)

The order in which the planning documents are developed is relatively unimportant, with two exceptions: the WBS must be the first document developed and approved, and the risk management plan should be the last planning document developed. The WBS is the basis from which all the other planning documents are developed; therefore, it must be developed first. Risks and opportunities are identified by examining the WBS and other planning documents. Hence, the risk management plan is the last of the planning documents to be developed and approved.

Sometimes the process of developing the planning documents is not entirely sequential. Sometimes the process is interactive and iterative. Decisions on one planning document may require changes to another planning document. Often, the first version of a planning document, e.g., the network diagram, needs to be tentative and changed later because of the availability of human resources being finalized in the resource plan.

The planning process is complete when the project team has the full complement of planning documents and the details of the planning documents are consistent and congruent with the other planning documents.

Role of the Project Manager

The project manager is expected to have the following:

1. The political awareness to build support for the project among important stakeholders and sponsors plus the ability to negotiate with functional managers for the assistance the project needs
2. The skills, tools, and procedures with which to lead the team in its efforts to accomplish the project activities
3. The people skills to build a committed and motivated team
4. The attitude to reinforce management priorities
5. The assessment skills to assemble a team of people with complementary skills and knowledge

Outputs of the Planning Phase

The outputs of the planning phase include the WBS, the budget, the project schedule, the resource plan, the communications plan, and the risk management plan. The planning phase seeks to achieve a condition where stakeholders and sponsors are comfortable with the way the project is progressing. Another desired outcome is a project team consisting of committed and motivated people. By the end of the planning phase, the team should have a strategy in place to facilitate the acceptance of the project deliverables by the important stakeholders, sponsors, and constituencies. This strategy is embedded in the communications plan.

Planning Phase Problems

The following problems may occur during this phase:

1. The business case definition has not been developed. The project team does not have any real understanding of what the sponsors see as the problems or needs that must be addressed. The team does not know the extent to which their vision of the project (objectives, assumptions, constraints, etc.) agrees with that of the sponsors or upper management. Under this circumstance

the first team activity should be to develop the business case definition, brief it, get it clarified and approved.

2. There is pressure being exerted upon the project manager to get the project going. "We don't have time for all that fancy and useless planning stuff! You know what you have to do, just do it!" This is a tough problem. You know you have to plan the project to execute and control it properly, but defying your sponsors or upper management is not advisable. Ask your team to work a weekend to develop a plan sufficient to determine what work packages are to be started first. Get work packages started to show upper management the movement they seek but, at the same time, ask the team to put in the extra hours to develop the WBS, network diagram, and budget documents. Brief these to upper management, if you think this appropriate, and complete the remainder of the planning documents as soon as possible. Make sure you show your appreciation for the team's extra effort at this crucial time. Pay for lunch and drinks after-hours when they work extra hours or weekends. Remember this extra effort when it's time for performance appraisals.

3. Upper management is not able to articulate clearly what it wants. This lack of clarity may exist at the total project level or at lower levels of work. The strategy here is to fill the void with a proposed statement that managers may react to, modify, and then approve. This is easier than presenting a "blank sheet of paper" and asking upper managers to draft the requirements. This strategy is really the basis for briefing the business case definition; it presents information and asks managers to react to it.

4. The norms of the organization accept sloppy estimates of time or cost to perform work. People from whom you seek estimates consider estimating an intrusion upon their legitimate work; they give you the "go away" estimate. Your request for an estimate is met with a quick, haphazard guess at an estimate, implying "now go away." In this case, the project manager should make a formal request for an estimate. Use a standardized estimating sheet similar to the one in Figure 10.4; ask the estimator and the estimator's supervisor to sign it. You may not make any friends this way but you will definitely indicate the professionalism you expect. If necessary, go to the sponsor (a powerful sponsor helps here) and ask for help in emphasizing the importance of good estimates. Sloppy estimates will make controlling the project a nightmare!

Chapter 4

Project Scope

Mega Recipe for Determining Project Scope

Activity 1: Review the existing scope documents.
Activity 2: Develop the work breakdown structure.
Activity 3: Write the work package work orders.

The scope of a project refers to the sum of all the work that must be performed in the project. Work within the scope of the project is work that will be performed, and work outside the scope is work that will not be performed. For example, there is a project to develop a new software application for a customer organization. A legitimate question concerns whether training the employees how to use the software is part of the project's scope. The project manager would be responsible to develop and conduct training to end users if it was previously decided that training is within the project's scope. Determining a project's scope, i.e., its final deliverables, is done as part of the project business case definition process during the initiation phase.

On any particular project, there may be a number of documents that discuss the project scope, i.e., indicating what deliverables the project must produce. Scope in this context refers to the final deliverables the project must produce and submit to the customer. Project business case definition, mission statement, charter, statement of work, and contract are some documents that may indicate the final deliverables the project must produce. The document that provides the definitive statement of the project's scope is called the work breakdown structure (WBS). Simply stated, the WBS is an organized list of all the work that must be performed in the project. (With reference to the WBS, scope means all the work that must be performed, which in turn will produce all the necessary final deliverables.) The WBS is called an organized list because the elements of work are listed under categories of work. The smallest element of work listed has three names: "work package," "task," or "activity." These names are used interchangeably although "work package" is the name most frequently used on formal projects such as building a ship, and "task" is often used on small informal projects such as planning a picnic. Guidelines and definitions are listed as follows:

1. A work package is an element of work that may be completed in 40 to 80 hours of effort, i.e., it may take from one to two weeks to complete. Sometimes a work package may require more than one person to perform. For example, "Frame the House" is a work package that is do-able within two weeks; it requires a crew of three or four carpenters. On small projects requiring a few weeks to complete, the work packages may be as small as one or two days of effort. However, the reverse is not true. On large, long projects such as building an aircraft carrier (which may take about 10 years to build), work packages are still elements of work that can be accomplished in 40 to 80 hours. Building an aircraft carrier is estimated to have about 350,000 work packages. In this case, there is not one WBS for the entire ship but many: one for each of the major components such as the engines and drive system, the radar system, the elevator systems, etc. Each may be built by a different company. There would be one top-level WBS that lists the work to assemble these major components into the total ship. The work packages on the top-level WBS would take from 40 to 80 hours to complete also.

2. Work package titles are shown in the WBS. Each work package title starts with a verb and ends with a noun. For example, "Develop Training Materials": "develop" is the verb and "materials" is the noun. Each work package produces a deliverable of some kind; in the example, the deliverable is the training material.

3. Each work package is assigned to a person, organization, or department to perform. The responsibility for completing a work package must not be bifurcated. For example, software design that is done in Atlanta can't be in the same work package title as software design performed in Philadelphia. The work done in each city must have its own work package title.

4. The WBS will consist of work package and category titles. The detailed information about each work package is included in the work package work order document. The work order is described later in this chapter. The format for a work order is shown in Figure 4.1.

5. On the WBS, the elements of work (work packages) are listed under categories. In Tables 4.1 through 4.4, categories are shown in bold type.

6. The titles of the categories and the project title should be nouns or noun phrases, e.g., "Engineering Department," "Installation," and "Pre-Move Activities." Tables 4.1 through 4.4 provide examples.

7. It is not necessary to list work packages in chronological order on the WBS. The network diagram is the document that shows the work packages in their chronological order.

Work Package Manager _____ Work Package Title _____ Work Package No._____

Describe scope of the work package _____ _____See Attachment

Description of Deliverable (s) _____ _____See Attachment

Budget Baseline _____ Max Acceptable Cost _____ Min Acceptable Cost _____

Effort Time Baseline _____ (Hours ____) (Days ____) (Weeks ____)

Max Acceptable Completion Time _____ Min Acceptable Completion Time _____

Early Start Date _____ Early Finish Date _____ *Total Float _____

Latest Start Date _____ Latest Finish Date _____

* None of the total float may be used without the Project Manager's prior approval.

Potential Problems or Risks _____ _____See Attachment

Acceptance Criteria (Describe how performance will be evaluated). Other (Describe): _____ _____See Attachment

_____ Acceptable deliverables

_____ Cost with Max and Min Control Limits

_____ Performance Time within Max and Min Control Limits

Figure 4.1 Work package work order form.

Table 4.1   WBS to Rebuild an Antique Car Developed via the Deliverables Approach

RESTORE ANTIQUE CAR

1.0   Project Management
    1.1   Develop project plan
    1.2   Monitor and control project
    1.3   Conduct closeout activities
2.0   Car Assembly
2.1   Wheels and Steering Assembly Installation
    2.1.1   Purchase new wheels and hubcaps
    2.1.2   Repair, repaint, grease axles
    2.1.3   Disassemble, repair, reassemble steering assembly
    2.1.4   Disassemble, repair, reassemble front and rear axles
2.2   Engine Installation
    2.2.1   Disassemble, clean, repair, reassemble engine
    2.2.2   Disassemble, clean, repair, reassemble fuel pump
    2.2.3   Disassemble, clean, repair, reassemble cam assembly
2.3   Transmission Installation
    2.3.1   Disassemble, clean, repair, reassemble transmission
2.4   Radiator and Coolant System Installation
    2.4.1   Clean, repair, paint radiator
    2.4.2   Replace hoses, gaskets, sealant, and coolant
2.5   Air Conditioner Installation
    2.5.1   Disassemble, repair, reassemble air conditioner
2.6   Interior Work
    2.6.1   Install new seats and interior carpeting
2.7   Body Exterior and Bumpers Installation
    2.7.1   Repaint car body and bumpers
    2.7.2   Install exterior lights and fixtures
2.8   Alternator System Installation
    2.8.1   Disassemble, clean, repair, reassemble alternator
2.9   Drive Shaft Assembly Installation
    2.9.1   Disassemble, repair, reassemble shaft assembly
2.10 Electrical System Installation
    2.10.1   Replace old wiring
    2.10.2   Replace battery

Note: Table 4.5 is the template for this type of WBS.

Table 4.2   WBS for Construction of a House Developed via the Function and Department Approach

HOUSE AT 2900 SUNRISE COURT

1.0   General Contractor (Project Manager)
    1.1   Get financing
    1.2   Get building and other permits
    1.3   Monitor and control project
    1.4   Conduct walk-through with customer

2.0   Land Developer
    2.1   Clear lot of trees and boulders
    2.2   Excavate the hole for basement
    2.3   Complete final grading of lot
    2.4   Landscape the lot

3.0   Masonry Contractor
    3.1   Build basement walls
    3.2   Build fireplace and chimney
    3.3   Pour walkways
    3.4   Pour driveway
    3.5   Complete brick veneer of outer walls
    3.6   Install tile in bathrooms

4.0   Carpentry Contractor
    4.1   Build frame of house and cover with plywood
    4.2   Shingle the roof
    4.3   Install interior wall studs and stairs
    4.4   Install wallboard to interior walls and ceilings
    4.5   Spackle and sand the interior walls and ceilings
    4.6   Install hardwood floors
    4.7   Install interior trim and doors

5.0   Plumbing Contractor
    5.1   Install rough plumbing lines
    5.2   Install fixtures and tubs
    5.3   Install appliances
    5.4   Install furnace and air conditioner

6.0   Painting Contractor
    6.1   Paint outside trim
    6.2   Paint inside walls
    6.3   Paint inside trim

Note: Table 4.7 is the template for this type of WBS.

Table 4.3 WBS for Production Control System Developed via the Separate Deliverables Approach

PRODUCTION CONTROL SYSTEM

1.0 Project Management
    1.1 Develop project plan
    1.2 Monitor and control the project
    1.3 Conduct closeout activities
    1.4 Purchase hardware

2.0 Production Control Software
    2.1 Determine current control system
    2.2 Develop software requirements
    2.3 Develop prototype of software
    2.4 Test prototype
    2.5 Modify software
    2.6 Demonstrate software

3.0 Installation
    3.1 Install software in engineering office
    3.2 Install computers in warehouses
    3.3 Install computers in production area
    3.4 Audit warehouse inventory records
    3.5 Create inventory database
    3.6 Test the system

4.0 Training
    4.1 Develop training materials
    4.2 Develop performance aids
    4.3 Train employees

Note: The deliverables are software system, installation of the system, and training of employees. Table 4.6 is the template for this type of WBS.

Activity 1: Review Project Scope Documents

Review all the existing documents that contain project scope information. These may include the proposal, contract, charter, business case definition, and others. Review these to assure that the team is aware of all the deliverables required of the project and other information that may impact the work that must be included in the WBS. For example, a review of the business case definition revealed that physical security of building materials must be provided by the project team. This requires a work package to rent chain-link fence and another to install the security fence.

Table 4.4   WBS for Circus Move Developed via Time Phase Approach

---

CIRCUS MOVE TO PLEASANTVILLE

1.0   Pre-Move Activities
    1.1   Determine site location in next town
    1.2   Contract for site use
    1.3   Get next town permits
    1.4   Determine water/sewer connection facilities at Pleasantville site
    1.5   Plan for circus and carnival attraction locations
    1.6   Plan for day-one rest area site location
    1.7   Buy advance food and other supplies
    1.8   Perform advance advertising and promotion

2.0   Shutting Down Activities (at current site)
    2.1   Disconnect electrical/water/sewer lines
    2.2   Secure animals in trailers
    2.3   Break down tents and attractions
    2.4   Pack tents and attractions onto trailers
    2.5   Fuel up all vehicles
    2.6   Feed and water animals
    2.7   Feed circus troop
    2.8   Shut down and pack kitchen

3.0   Travel Activities
    3.1   Transport vehicles for day one
    3.2   Feed and exercise animals at day-one rest area
    3.3   Set up field kitchen
    3.4   Feed and overnight rest for the circus troop
    3.5   Feed animals
    3.6   Feed circus troop breakfast
    3.7   Shut down and pack kitchen
    3.8   Transport vehicles to Pleasantville

4.0   Setting Up Activities at Pleasantville
    4.1   Set up animal housing area
    4.2   Set up tents
    4.3   Set up attractions
    4.4   Feed and exercise animals
    4.5   Make electrical/water/sewer connections
    4.6   Set up field kitchen
    4.7   Feed circus troop
    4.8   Open the circus and attractions for business

Note: See Table 4.8 for format.

Activity 2: Develop the Work Breakdown Structure (WBS)

There are three ways to develop the WBS (discussed below). It is the nature of the project's final deliverables and the relationship among work packages that determine which method to use.

Listing by Deliverables

This method is appropriate when the project's purpose is to produce one or more clearly delineated deliverables. Each major deliverable is broken down into sub-deliverables and work packages; see Tables 4.1 and 4.3. In Table 4.1, the restored car is the total project deliverable. The major deliverables include project management and the assembled car. The major sub-deliverables include body, engine, steering and wheel assembly, transmission, air conditioner, radiator coolant system, alternator, drive shaft assembly and the electrical system. Each of these major sub-deliverables is further broken down into work packages. The by-deliverables approach uses this deductive method to break the work down from the total deliverable (the car) into smaller and smaller elements of work until you arrive at elements of work that can be performed within 40 to 80 hours of effort; this is the work package level. Table 4.5 provides the format for developing this type of WBS.

The by-deliverables approach is also useful when the project consists of a number of separate and distinct deliverables that together constitute the project final deliverables. Each deliverable is broken into work packages. Table 4.3 is an example; in this WBS, the final deliverables are project management, production control software, installation, and training. Table 4.6 provides the template for developing this type of WBS.

Listing by Function or Department

This approach to developing a WBS is appropriate when the major project deliverable is produced through the efforts (work packages) of numerous functions or departments. The major division of work is by department or function. Table 4.2 is an example. In this WBS, general contractor, masonry contractor, and carpentry contractor are names of contractors who perform a function on the project. Table 4.7 is the template for this type of WBS.

Listing by Time Phase

This approach is appropriate when the project clearly breaks down into groups of activities (tasks) to be performed by time phase. The "Circus Move to Pleasantville" (Table 4.4) is an example of a project where certain groups of tasks must be

Table 4.5 Four-Level "By Deliverable" WBS Format

Level 1[a]	**Total Project** (Note 1)		
Level 2	**1.0**	**Project Management**	
Level 4		1.1	(Note 2)
		1.2	(Note 2)
		1.3	(Note 2)
		1.4	(Note 2)
Level 2	**2.0**	**Major Deliverable #1** (Note 3)	
Level 3		2.1	Sub-deliverable #1 (Note 4)
Level 4			2.1.1 Work Package #1 (Note 5)
			2.1.2 Work Package #2 (Note 5)
			2.1.3 Work Package #3 (Note 5)
			2.1.4 Work Package #4 (Note 5)
Level 3		2.2	Sub-deliverable #2 of (Note 6)
Level 4			2.2.1 Work Package #1 (Note 7)
			2.2.2 Work Package #2 (Note 7)
			2.2.3 Work Package #3 (Note 7)
			2.2.4 Work Package #4 (Note 7)
Level 3		2.3	Sub-deliverable #3 (Note 8)
Level 4			2.3.1 Work Package #1 (Note 9)
			2.3.2 Work Package #2 (Note 9)
			2.3.3 Work Package #3 (Note 9)
			2.3.4 Work Package #4 (Note 9)

[a] Level 1 is the name of the project. Level 2 includes the category names for major pieces of work (e.g., project management or names of major deliverables). Level 3 lists the names of sub-deliverables needed to produce the major deliverable. The work packages are listed under each sub-deliverable. This format is useful when the purpose of the project is to produce a major deliverable (e.g., hardware) that is composed of sub-deliverables (sub-assemblies). The format shows one major deliverable made up of three sub-deliverables.

Notes: (1) Indicate the name of the project here. (2) Indicate the names of the work packages under the project management category; e.g., develop project plan, monitor, and control the project, conduct closeout activities, etc. (3) Indicate the name of the first major deliverable here. In the format above, one major deliverable is shown and it is composed of three sub-deliverables. Add additional major deliverables if necessary. (4) Indicate the name of sub-deliverable #1. (5) Indicate the names of the work packages under sub-deliverable #1. The format shows four work packages needed to produce sub-deliverable #1. Add or subtract work packages as necessary. (6) Indicate the name of sub-deliverable #2. (7) Indicate the names of the work packages needed to produce sub-deliverable #2. The format shows four work packages needed; add or subtract work packages as necessary. (8) Indicate the name of sub-deliverable #3. (9) Indicate the names of the work packages needed to produce sub-deliverable #3. (10) Extend this format, adding category names (major and sub-deliverable names) and work packages as necessary. It is not necessary to indicate level number on an actual WBS.

Table 4.6   Three-Level "By Deliverable" WBS Format

Level 1[a]	**Total Project** (Note 1)	
Level 2	**1.0**	**Project Management**
Level 3		1.1 (Note 2)
		1.2 (Note 2)
		1.3 (Note 2)
		1.4 (Note 2)
		1.5 (Note 2)
Level 2	**2.0**	**Deliverable #1** (Note 3)
Level 3		2.1 (Note 4)
		2.2 (Note 4)
		2.3 (Note 4)
		2.4 (Note 4)
		2.5 (Note 4)
Level 2	**3.0**	**Deliverable #2** (Note 5)
Level 3		3.1 (Note 6)
		3.2 (Note 6)
		3.3 (Note 6)
		3.4 (Note 6)
		3.5 (Note 6)
Level 2	**4.0**	**Deliverable #3** (Note 7)
Level 3		4.1 (Note 8)
		4.2 (Note 8)
		4.3 (Note 8)
		4.4 (Note 8)
		4.5 (Note 8)

[a] Level 1 is the name of the project. Level 2 includes the category names for major pieces of work (e.g., project management) or names of major deliverables. Level 3 lists the names of work packages under each category. See Table 4.3 for an example. This format is useful when the purpose of the project is to produce separate and distinct deliverables. The WBS template for a project to produce a single deliverable and its sub-deliverables is shown in Table 4.5.

Notes: (1) Indicate the name of the project here. (2) Indicate the names of the work packages under the project management category; e.g., develop project plan, monitor and control the project, conduct closeout activities, etc. (3) Indicate the name of the first major deliverable here. (4) Indicate the names of the work packages needed to produce major deliverable #1. (5) Indicate the name of the second major deliverable here. (6) Indicate the names of the work packages needed to produce major deliverable #2. (7) Indicate the name of the third major deliverable here. (8) Indicate the names of the work packages needed to project major deliverable #3. (9) Extend this format, adding categories (deliverable names) and work packages as necessary. It is not necessary to indicate level number on an actual WBS.

Table 4.7 WBS Format Based on Contributing Functions and Departments

Level 1[a]	**Total Project** (Note 1)	
Level 2	**1.0**	**Project Management**
Level 3		1.1 (Note 2)
		1.2 (Note 2)
		1.3 (Note 2)
		1.4 (Note 2)
Level 2	**2.0**	**Function/Department 1** (Note 3)
Level 3		2.1 (Note 4)
		2.2 (Note 4)
		2.3 (Note 4)
		2.4 (Note 4)
Level 2	**3.0**	**Function/Department 2** (Note 5)
Level 3		3.1 (Note 6)
		3.2 (Note 6)
		3.3 (Note 6)
		3.4 (Note 6)
Level 2	**4.0**	**Function/Department 3** (Note 7)
Level 3		4.1 (Note 8)
		4.2 (Note 8)
		4.3 (Note 8)
		4.4 (Note 8)

[a] Level 1 is the name of the project. Level 2 includes the names of the departments or functions of the organization supporting the project; see Table 4.2 for an example. Level 3 lists the names of work packages under each department or function. Use this format to list departments' functions, or contractors participating in the project. List work packages under the names of the departments or functions. Each category (department, function, or contractor) has its own budget.

Notes: (1) Indicate the name of the project here. (2) Indicate the names of the work packages under the project management function, e.g., develop project plan, monitor and control the project, conduct closeout activities, etc. (3) Indicate here the name of a department supporting the project, e.g., training department. (4) Indicate the names of the work packages to be performed by the department 1 indicated in (3): determine training needs, develop training objectives, develop training material, reproduce training materials, and conduct training. (5) Indicate the name of a second department or function supporting the project, e.g., Installation and Testing. (6) Indicate the names of the work packages to be performed by the second department or function: develop the test protocol, install the system, test the system, fix problems. (7) Indicate the name of the third department or function supporting the project. (8) Indicate the names of the work packages to be performed by the third department or function. (9) Extend this format, adding category names (department and functions) and work packages as necessary. It is not necessary to indicate level number on an actual WBS.

performed before the main event (the move). Some tasks must be performed during the move and others must be performed after the move. The project has a clear overarching time structure: before-move events, move events, and after-move events. There is little connection among the tasks within a time phase other than they are performed in the same time phase. Within each category (shown in bold type), the tasks are not in chronological order. Using the by-deliverable approach and the by-function or by-department approaches are not useful for this kind of project. Table 4.8 provides the template for this type of WBS.

Assigning numbers to the WBS: Numbers need to be added for each category and every work package. A common way to do this is to assign a number starting with 1.0 to each category and assign a number for each work package using the tenths digit. For example, in Table 4.4, the first category, "Pre-Move Activities," is account 1.0. The other categories are numbered 2.0 through 4.0. Eight work packages are listed under category 1.0, numbered 1.1 through 1.8.

WBS approval: After the WBS is completed, it must be reviewed by the entire team and approved by all stakeholders because the WBS is the foundation upon which the rest of the project plan is built. It is the document from which the budget, network diagram, Gantt chart, resource plan, and risk management plan are built. Any change to the WBS will have consequent impacts upon these other planning documents. Make sure the WBS lists all the work packages needed to produce all the required deliverables including those needed to close out the project. Get it approved before proceeding to the other planning documents.

Activity 3: Write the Work Package Work Orders

Although the WBS lists all the work that must be accomplished in the project, it does not provide the detailed information needed by the person or department that will perform the work package. The detailed information for each work package is included on a form called the work package work order. Figure 4.1 is the format. It is initiated after the WBS is developed but before it is approved. The work order is needed to supplement the WBS because the WBS only lists work package titles. The WBS does not provide information about the scope of each work package, the subtasks required of the work package, or the deliverables that must be produced from the work package. After the WBS is completed, the "Describe Scope of the Work Package" and the "Description of Deliverable(s)" portions of the work order must be completed. The WBS is then submitted for approval along with the work orders for all the work packages. Together, these documents provide the detailed and comprehensive definition of project scope. The remaining parts of the work order form will be completed during the planning phase. A copy will be given to the person (or department or vendor) responsible for performing the work package.

Table 4.8   Format for a WBS Developed via Time Phase Approach[a]

Level 1	**Project Name** (Note 1)	
Level 2	**1.0 Time Phase 1** (Note 2)	
Level 3		1.1 (Note 3)
		1.2 (Note 3)
		1.3 (Note 3)
		1.4 (Note 3)
Level 2	**2.0 Time Phase 2** (Note 4)	
Level 3		2.1 (Note 5)
		2.2 (Note 5)
		2.3 (Note 5)
		2.4 (Note 5)
Level 2	**3.0 Time Phase 3** (Note 6)	
Level 3		3.1 (Note 7)
		3.2 (Note 7)
		3.3 (Note 7)
		3.4 (Note 7)
Level 2	**4.0 Time Phase 4** (Note 8)	
Level 3		4.1 (Note 9)
		4.2 (Note 9)
		4.3 (Note 9)
		4.4 (Note 9)

[a] This format is useful when the project clearly breaks down into time sequenced groups of activities where phase 1 activities must be performed before phase 2 activities, phase 2 activities before phrase 3 activities, etc. There is little connection among the tasks within a time phase other than they are performed in the same time phase. The "by deliverable" and the "by function or department" approaches are not very useful for a time phased project like moving a group of employees to a new location or moving the circus to the next town, as in Table 4.4.

Notes: (1) Indicate the name of the project here. (2) Give a name to the category of tasks that must be performed in the first phase of the project. "Pre-Move Planning Activities" was the name given to the first set of activities in the project to move the circus in Table 4.4. (3) Indicate the names of the tasks to be performed in the first phase of activities. (4) Indicate a name to the category of tasks that must be performed in the second phase of the project. "Shutting Down Activities" was the name given to this second set of activities in Table 4.4. (5) Indicate the names of the tasks to be performed in the second phase. (6) Give a name to the category of tasks that must be performed in the third phase of the project. "Travel Activities" was the name given to this third set of activities in Table 4.4. (7) Indicate the names of the tasks to be performed in the third phase. (8) Give a name to the category of tasks that must be performed in the last phase of the project. "Setting Up Activities" was the name given to this last set of activities in Table 4.4. (9) Indicate the names of the tasks to be performed in the last phase. (10) Modify this format to match the project's flow of work. It is not necessary to indicate level numbers on the actual WBS.

Notes

1. For many years, category names were referred to as cost accounts. The name "cost account" comes from the fact that each category had its own budget, which was an account of money. For example, Engineering Department may be the name of a cost account. It has a budget (account) of $25,000 to perform its work packages. Cost account is being replaced by the term "control account" because the category name may be the point where performance is measured.

Chapter 5

Project Cost

Mega Recipe for Project Cost

Activity 1: Determine work package cost.
Activity 2: Determine total project cost.

Role of the Project Manager

The project manager's primary role is to encourage, facilitate, and insist upon accurate estimates of cost. The project manager does this by providing procedures and formats, and requiring formal documentation of the estimating process.

Output of the Cost Estimating Activity

The project manager and management have confidence in the cost estimates.

Activity 1: Determine Work Package Cost

Work package cost consists of material cost, equipment cost, and labor cost. If a work package has material and equipment costs, these must be included in the estimate.

> **Example 1:** For the work package "Install Sewer and Water Lines," this includes the labor and equipment cost to dig the trench, the cost of stone gravel placed in the trench, the cost of the copper tubing that will transport water, and the cost of the sewer pipe. The quantities of

these materials are estimated from previous projects using the analogy or parametric techniques (discussed later in the chapter). The estimates: 4 cubic yards of gravel, 75 feet of sewer pipe, and 100 feet of copper tubing. Our supplier provides the material prices: $50 per cubic yard (cu yd) for gravel, $2 per lineal foot (ln ft) for sewer pipe, and $1 per lineal foot for copper tubinge.

Material	Estimating Equation		Cost
Gravel	4 cu yd × $50/cu yd	=	$200
Sewer pipe	75 ln ft × $2/ln ft	=	$150
Copper tubing	100 ln ft × $1/ln ft	=	$100
Total Material Cost		=	$450

Labor costs are determined from estimates of how long each labor category will need to accomplish its part of the work package. From a previous project, we estimate that the backhoe operator will need four hours to dig and refill the trench. Likewise, we estimate from a previous project that the plumber will need eight hours to connect the water and sewer pipelines. The loaded labor rates (LLR), the cost per hour for their time for the backhoe operator and plumber are $120 and $100 per hour, respectively.

Labor Category	Estimating Equation		Cost
Backhoe operator	4 hours × $120/hour	=	$480
Plumber	8 hours × $100/hour	=	$800
Total labor cost		=	$1,280

The equipment cost for this work package includes the rental cost of the backhoe machine. From a previous project it is estimated that the backhoe will be required for a day. Cost per day is $500.

Equipment	Estimating Equation		Cost
Backhoe rental	1 day × $500/day	=	$500
Total equipment cost		=	$500

The estimate of total cost for this work package is the sum of the material, labor, and equipment costs: $2,230. Table 5.1 shows how this information is documented on the work package estimating sheet. Table 5.2 is the template.

Table 5.1   Example of Estimating Work Package Total Cost Using Single Estimates of Effort Time for Each Task

WORK PACKAGE ESTIMATING SHEET

Project Name: House#24   **Work Package:** 5.1 Install water/sewer lines
Project Manager: JBR   **Date:** 10/3/08
Work Package Description: Install water and sewer lines from street connection points to water meter and first sewer trap in basement — approximately 75 feet.
Assumptions/Constraints/Risks: None

Labor Cost

Tasks/Deliverables	#People	LLR	Effort Time	Cost
Dig trench w/backhoe	1	$120	4 hours	$480
Lay gravel, install water line and sewer line	1	$100	8 hours	$800
			Total Labor Cost	**$1280**

Materials Cost

Description of Materials	Quantity	Unit Cost	Item Cost
Gravel stone	4 cu yards	$50/cu yd	$200
Sewer pipe	75 feet	$2.00/ft	$150
Copper tubing	100 feet	$1.00/ft	$100
		Total Materials Cost	**$450**

Equipment Cost

Description of Equipment	Quantity	Unit Cost	Item Cost
Backhoe rental	1 day	$500/day	$500
		Total Equipment Cost	**$500**
		Total Work Package Cost	$2,230

Notes: (1) Table 5.2 is the template for this estimate. (2) Table 5.3 shows an example using the PERT method. (3) LLR refers to the loaded labor rate, which is the dollar cost for each hour of labor.

Adding Risk to the Estimates[1]

The method described above does not include risk; the estimates of labor costs are at the unknown level of confidence, probably close to 50%. The largest risk associated with the work package cost is associated with the labor cost. PERT, a series of simple equations, provides the means to determine the work package cost at specific levels of confidence. The PERT approach requires three estimates of how long it will take to do each task: the pessimistic estimate (P) is the longest it could take, the optimistic estimate (Op) is the shortest it could take, and the most likely (ML) is the estimate when the task does not experience the worst conditions nor does it experience the best conditions.

Table 5.2 Template for Estimating Work Package Total Cost Using a Single Estimate of Effort Time for Each Task

WORK PACKAGE ESTIMATING SHEET	

Project Name: Work Package:
Project Manager: Date:
Work Package Description:
Assumptions/Constraints/Risks:

Labor Cost

Tasks/Deliverables	# People	×	LLR	×	Effort Time	=	Cost

Total Labor Cost: $

Materials Cost

Description of Materials	Quantity	×	Unit Cost	=	Item Cost

Total Materials Cost: $

Equipment Cost

Description of Equipment	Quantity	×	Unit Cost	=	Item Cost

Total Equipment Cost: $
Total Work Package Cost $

Instructions: (1) Enter the project name, work package title, project manager, and date. (2) Enter the description of the work package. Refer to the work package work order and attach a copy. (3) Indicate any assumptions, constraints, or risks associated with the work package. They may be documented on the work order. **Under "Labor Cost":** (4) List the tasks or deliverables that must be produced by the work package. (5) Indicate the number of people who will work on each task or deliverable. (6) Indicate the loaded labor rate (LLR) for each person working on each task or deliverable. If two people are to work on a task, indicate their combined LLR. Indicate the combined LLR with the letter "t" after the rate. For example, two people are to work on a task; one has an LLR of $100/hour, the other a rate of $120/hour. Indicate the combined LLR as $220t. (7) Indicate the estimated time to complete each task or deliverable in the "Effort Time" column. If the task is a two-person task and requires that they work together for 40 hours, indicate 40 hours in the "Effort Time" column. Always enter the units of time: hours or days. (8) For the two-person team, multiply the combined LCR ($220t in our example) and effort time; enter this figure in the "Cost" column. (9) Sum the figures in the "Cost" column; enter this as total labor cost. **Under "Materials Cost":** (10) List the materials, quantity, item cost, for all the materials needed to perform the work package. (11) Sum the item costs and indicate this figure as total materials cost. **Under "Equipment Cost":** (12) List the equipment, quantity, item cost, for all tools and equipment. (13) Sum the item costs and indicate this figure as total equipment cost. **Under "Total Work Package Cost":** (14) Sum the total labor cost, total material cost, and total equipment cost. Indicate this figure as the total work package cost.

Note: Table 5.4 shows the template for using the PERT method.

We get three estimates of duration time from the backhoe operator and the plumber. For the backhoe work, the optimistic estimate (least time it could take) is three hours, the pessimistic time (longest time it could take) is eight hours, and the most likely time is four hours. For the plumber, the optimistic time is six hours, the pessimistic time is 16 hours, and the most likely time is eight hours.

PERT Equations

$$\text{Average effort time (Te)} = [P + 4(ML) + Op] \div 6 \qquad (5.1)$$

$$\text{Standard deviation } (\sigma) = [P - Op] \div 6 \qquad (5.2)$$

$$\text{Maximum duration time}_{.95} = Te + 2\sigma \qquad (5.3)^3$$

$$\text{Minimum duration time}_{.95} = Te - 2\sigma \qquad (5.4)$$

Backhoe Operator	Plumber

Average time = [8 + 4(4) + 3] ÷ 6
  = 4.5 hours
	Average time = [16 + 4(8) + 6] ÷ 6
	  = 9 hours

Standard deviation = [8 – 3] ÷ 6
  = .8 hours
	Standard deviation = [16 – 6] ÷ 6
	  = 1.7 hours

Maximum duration time.97
  = 4.5 + 2(.8) = 6.2 hours
	Maximum duration time.97 = 9 + 2(1.7)
	  = 12.4 hours

Average labor cost = 4.5 hours ×
  $120/hour = $540
	Average labor cost = 9 hours ×
	  $100/hour = $900

Cost at 97% level of confidence
  = 6.1 hours × $120/hour = $732
	Cost at 97% level of confidence
	  = 12.4 hours × $100/hour = $1,240

For the backhoe operator, the average time to dig and refill the trench is 4.5 hours, and we are 95% confident that this work can be completed (under the worst conditions) in 6.2 hours. (Round all hours and standard deviation numbers to the nearest tenth; round dollar amounts to the nearest dollar.) With this, we can determine the average cost and the cost at 95% confidence level.

Table 5.3 shows how the PERT approach has been applied to the labor cost and only the labor cost part of the work package estimate. We have three estimates of the work package cost. The only difference among the three estimates is how labor cost was determined. The first method used a single estimate of how long it would take to complete the work package; this produced the one-point estimate cost of $2,230 (see Table 5.1). The PERT equations provided two estimates of work package cost: the average cost of $2,390 and the 95% confidence level cost of $2,923. The PERT method is shown in Table 5.3; Table 5.4 is the template for this method. (The single-point method is the easiest but it may seriously underestimate the cost and you don't know how much confidence to place in the estimate.) The PERT methods give estimates at 50% and 95% levels of confidence; these require a little more work. You decide which method to use.[2]

Table 5.3 Example of Estimating Work Package Total Cost Using PERT Estimates of Effort Time for Each Task

WORK PACKAGE ESTIMATING SHEET

Project Name: House # 24
Project Manager: JBR

Work Package: 5.1 Install water/sewer lines
Date: 10/3/08

Work Package Description: Install water and sewer lines from street connection points to water meter and first sewer trap in basement — approximately 75 feet.
Assumptions/Constraints/Risks: None

Labor Cost

Tasks/Deliverables	#People	LLR	Op	ML	P	Avg Time	Std Dev	Avg Cost	.95 Time[3]	.95 Cost[3]
Dig trench w/backhoe	1	$120	3 hours	4	8	4.5	.8 hours	$540	6.1 hours	$732
Lay gravel, install water line and sewer line	1	$100	6 hours	8	16	9	1.7 hours	$900	12.4 hours	$1,240
Total Labor Cost								$1440	vs.	$1,972

Materials Cost

Description of Materials	Quantity	Unit Cost	Item Cost
Gravel stone	4 cu yards	$50/cu yd	$200
Sewer pipe	75 feet	$2.00/ft	$150
Copper tubing	100 feet	$1.00/ft	$100
Total Materials Cost			$450

Equipment Cost

Description of Equipment	Quantity	Unit Cost	Item Cost
Backhoe rental	1 day	$500/day	$500
Total Equipment Cost			$500

Total Work Package Cost $2,390 average vs. $2,922 at 95% confidence

Notes: (1) Table 5.1 shows an example using the single estimate method. This table shows that the PERT average estimated cost is $2,390 and the estimate at 95% confidence is $2,922. This table also shows how the PERT average effort times and effort times at 95% confidence are calculated. (2) LLR refers to the loaded labor rate, which is the dollar cost for each hour of labor. Op, ML, and P refer to the optimistic, most likely, and pessimistic estimates of effort time. Table 5.4 is the template for this estimate. (3) Remember to round average time and standard deviation to the nearest tenth and dollars to the nearest whole dollar.

Table 5.4 Template for Estimating Work Package Total Cost Using PERT Estimates of Effort Time for Each Task

WORK PACKAGE ESTIMATING SHEET

Project Name:
Project Manager:
Work Package Description:
Assumptions/Constraints/Risks:

Work Package:
Date:

Labor Cost

| Tasks/Deliverables | #People | Effort Time | | | Avg. Time | Std Dev | Avg. Cost | .95 Time[3] | .95 Cost[3] |
		LLR	Op	ML	P					

Total Labor Cost $ vs. $

Materials Cost

Description of Materials	Quantity	x	Unit Cost	=				Item Cost

Total Materials Cost $ vs. $

Equipment Cost

Description of Equipment	Quantity	x	Unit Cost	=				Item Cost

Total Equipment Cost $ vs. $

Total Work Package Cost   Average Cost: $ vs. $

Cost at 95% confidence

Table 5.4   Template for Estimating Work Package Total Cost Using PERT Estimates of Effort Time for Each Task (Continued)

Notes: (1) The .95 time and .95 cost are the maximum time and cost at 95% confidence. The minimum time and cost at 95% confidence are not shown. The maximum estimates at 95% confidence may be replaced with the time and cost at other levels of confidence; see instruction notes 7 and 8. (2) (LLR refers to the loaded labor rate, which is the dollar cost for each hour of labor. Op, ML, and P refer to the optimistic, most likely, and pessimistic estimates of effort time.)

Instructions: (1) Enter the project name, work package title, project manager, and date. (2) Enter the description of the work package. Refer to the work package work order and attach a copy. (3) Indicate any assumptions, constraints, or risks associated with the work package. They may be documented on the work order.

Under "Labor Cost": (4) List the tasks or deliverables that must be produced by the work package. (5) Indicate the number of people who will work on each task or deliverable. (6) Indicate the loaded labor rate (LLR) for each person working on each task or deliverable. If two people are to work on a task, indicate their combined LLR. Indicate the combined LLR with the letter "t" after the rate. For example, two people are to work on a task; one has an LLR of $100/hour, the other a rate of $120/hour. Indicate the combined LLR as $220t. (7) Enter the optimistic (Op), most likely (ML), and pessimistic (P) estimates of team effort time. Calculate the PERT average time, and standard deviation; enter this information. Multiply LLR and average time; enter this as the average cost. Add average time and two standard deviations to determine the .95 time. Enter this figure in the ".95 Time" column. Multiply .95 time and LLR to determine the .95 cost. Enter this figure in the ".95 Cost" column. Always enter the units of time: hours or days. Example: LLR is $50/hr; Op = 10; ML = 15; P = 25 hours; average time = [P + 4(ML) + Op] ÷ 6; average time = 15.8 hours; standard deviation = [P − Op] ÷ 6; standard deviation = 2.5 hours; average cost = average time × LLR; average cost = $790; .95 time = average time + 2 standard deviations; .95 time = 15.8 + 5 = 20.8 hours; .95 cost = .95 time × LLR; .95 cost = $1,040. Figure 5.1 is a spreadsheet template that performs these PERT calculations at the 50%, 90%, and 95% levels of confidence. (8) Sum average cost and enter this figure as the total average cost. Sum .95 cost and enter this as total .95 labor cost.

Under "Materials Cost": (9) List the materials, quantity, unit cost, and item cost for all the materials needed to perform the work package. (10) Sum the item costs and indicate this figure as total materials cost.

Under "Equipment Cost": (11) List the equipment, quantity, and item cost for all tools and equipment. (12) Sum the item costs and indicate this figure as total equipment cost.

Under "Total Work Package Cost": (13) Sum the total average labor cost, total material cost, and total equipment cost; enter this as the total work package average cost. Sum the .95 total labor cost, total material cost, and total equipment cost; enter this as the .95 total work package cost.

A	B	C	D	E	F	G	H	I	J	K	L	M	N
				Loaded Labor	Average 50%	Average 50%				90%	90%	95%[1]	95%
Category	Opt	ML	Pess	Rate/Hr	Eff Hrs	Cost	Std Dev			Eff Hrs	Cost	Eff Hrs	Cost
Category 1													
1.1					0.0	$0	0.00			0.00 $	-	0.00 $	-
1.2					0.0	$0	0.00			0.00 $	-	0.00 $	-
1.3					0.0	$0	0.00			0.00 $	-	0.00 $	-
1.4					0.0	$0	0.00			0.00 $	-	0.00 $	-
1.5					0.0	$0	0.00			0.00 $	-	0.00 $	-
1.6					0.0	$0	0.00			0.00 $	-	0.00 $	-
1.7					0.0	$0	0.00			0.00 $	-	0.00 $	-
1.8					0.0	$0	0.00			0.00 $	-	0.00 $	-
1.9					0.0	$0	0.00			0.00 $	-	0.00 $	-
Category 2													
2.1					0.0	$0	0.00			0.00 $	-	0.00 $	-
2.2					0.0	$0	0.00			0.00 $	-	0.00 $	-
2.3					0.0	$0	0.00			0.00 $	-	0.00 $	-
2.4					0.0	$0	0.00			0.00 $	-	0.00 $	-
2.5					0.0	$0	0.00			0.00 $	-	0.00 $	-
2.6					0.0	$0	0.00			0.00 $	-	0.00 $	-
2.7					0.0	$0	0.00			0.00 $	-	0.00 $	-
2.8					0.0	$0	0.00			0.00 $	-	0.00 $	-
2.9					0.0	$0	0.00			0.00 $	-	0.00 $	-

Figure 5.1 Excel© spreadsheet for estimating duration times and costs at 50%, 90%, and 95% levels of confidence.

Instructions: (1) Enter the category numbers and work package numbers in column A. Expand the spreadsheet as necessary. Get information from the WBS. (2) Enter the optimistic, most likely, and pessimistic estimates of effort time (in hours) in columns B, C, and D, respectively. (3) Enter the loaded labor rate for each work package in column E. If more than one person works on a work package, enter the sum of the loaded labor rates for all the workers. If necessary, change the rate and durations from hours to days. (4) For each work package, the spreadsheet will calculate the average (50%) effort time and average cost, and the effort time and cost at 90% and 95% levels of confidence. (5) Select the level of confidence desired (50%, 90%, or 95%) and use the effort time and cost at the selected level of confidence. (6) Enter the effort time and cost from step 5 into Table 5.3.

Activity 2: Determine Project Cost

Four methods are used for determining a project's cost. The first is the bottom-up technique, which is appropriate when you need the most accurate estimate — one that you will be held accountable to meet. The function/budget method is an alternative to the bottom-up technique because under firm fixed price circumstances it can be quite accurate. The analogy and parametric methods are less accurate methods suitable only during the initiation phase when a "ballpark" estimate is acceptable.

Bottom-Up Estimate

The bottom-up method is the most accurate technique because it requires that the work of the project be broken down into small, easily estimated pieces of work called work packages. The estimate from this method is accurate enough to serve as the project cost baseline.

The bottom-up method requires three steps:

1. Develop a complete work breakdown structure; the WBS must list every work package that is included in the project.
2. Estimate the cost of each work package using the methods described earlier in this chapter. Document the estimates on the work package estimating sheet.
3. Add the estimates of the work packages to determine the subtotals for each category and the total project cost.

Table 5.5 is an example. Table 5.6 is the template for the bottom-up estimate.

Functional Estimate

In this approach, the functional manager, department head, or contractor estimates the cost of each category of work. Each department participating in the project determines the work it will perform under a particular category (or it is listed in the WBS) and the equipment and material it will need, and estimates from this information the total cost of performing all the work within the category. The technique is also called the budget method because each department's estimate eventually becomes the department's budget. This method can be very accurate when the estimators are held accountable for the accuracy of their estimates such as firm fixed contracts in the following example. This method may not be very accurate when, for example, different departments, such as engineering department or software development, provide estimates and can then charge an amount in excess of the estimates.

> **Example 1:** The cost to produce a house is broken into the construction trade specialties: general contractor management, land development/landscaping, masonry, carpentry, electrical work, plumbing work, and painting. Each of these is a category of work to be performed by a construction subcontractor. See the WBS in Table 5.7.

Table 5.5 Example of Bottom-Up Estimate

PROJECT: INVENTORY CONTROL SOFTWARE SYSTEM

	Budget Estimate	
1.0 Project Management	**$16,000**	**Subtotal**
1.1 Develop project plan	5,000	
(4 people, 5 days, LLR = $250/day)		
1.2 Monitor and control project	6,000	
(1 person, 40 days half time, LLR = $300/day)		
1.3 Conduct closeout activities	5,000	
(4 people, 5 days, LLR = $250/day)		
2.0 Engineering/Design	**$34,800**	**Subtotal**
2.1 Develop system design	1,400	
(1 engineer, 4 days, LLR = $350/day)		
2.2 Develop test protocol	$700	
(1 engineer, 2 days, LLR = $350/day)		
2.3 Conduct system test	$700	
(1 engineer, 2 days, LLR = $350/day)		
2.4 Modify off-the-shelf software	$32,000	
(2 engineers, 30 days, LLR = 325/day		
software cost = $12,500)		
3.0 Installation	**$27,875**	**Subtotal**
3.1 Install hardware	$24,875	
(3 people, 15 days, LLR = $275/day		
hardware components cost = $12,500)		
3.2 Install software	$3,000	
(2 people, 5 days, LLR = $300/day)		
4.0 Training	**$1,750**	**Subtotal**
4.1 Develop training materials	$1,000	
(1 person, 5 days, LLR = $200/day)		
4.2 Conduct training	$750	
(1 person, 3 days, LLR = $250/day)		
	$80,425	**Total Project**

Notes: (1) Information in parentheses is not usually shown. (2) Table 5.6 is the template for the bottom-up estimate.

Successful contractors and subcontractors keep current and accurate data from which to estimate their costs. General contractors can use firm fixed price contracts with their subcontractors to reduce their risks. The general contractor has increased the accuracy of the estimate considerably and reduced cost risk because all costs, except carpentry and masonry materials and appliances, are firm fixed prices.

Table 5.6 Template for a Bottom-Up Estimate of Project Cost Based Upon a Three-Level WBS[a]

BOTTOM-UP ESTIMATE FOR		Note 1
1.0 Project Management	(Note 2)	**Subtotal**
1.1 (Note 3)		
1.2		
1.3		
1.4		
1.5		
2.0 Category 2 Name (Note 4)	(Note 4a)	**Subtotal**
2.1 (Note 5)		
2.2		
2.3		
2.4		
2.5		
3.0 Category 3 Name (Note 6)	(Note 6a)	**Subtotal**
3.1 (Note 7)		
3.2		
3.3		
3.4		
3.5		
4.0 Category 4 Name (Note 8)	(Note 8a)	**Subtotal**
4.1 (Note 9)		
4.2		
4.3		
4.4		
4.5		
5.0 Category 5 Name (Note 10)	(Note 10a)	**Subtotal**
5.1 (Note 11)		
5.2		
5.3		
5.4		
	(Note 12)	**Project Total**

[a] Use this template to create bottom-up estimates for projects similar to the examples in Tables 4.2 through 4.4, Table 5.5, and the templates in Tables 4.6 through 4.8. Extend this format, adding categories (deliverable names, function, or department names or time phase names) and work packages as necessary.

Notes: (1) In the space marked "Note 1" indicate the name of the project. (2) In the space marked "Note 2", enter the sum of the estimated costs of all the work

Table 5.6 Template for a Bottom-Up Estimate of Project Cost Based Upon a Three-Level WBS[a] (Continued)

packages listed under Project Management. (3) Enter the names of all the work packages to be performed under the project management category. In the right-most column, enter the estimated cost of each work package. The method of estimating work package cost is described in the "Work Package Cost" section of the text. (4) Replace "Category 2 Name" with the name of the second category. In the space marked "Note 4a", enter the sum of the estimated costs of all the work packages listed under the second category. (5) Enter the names of all the work packages to be performed under the second category. Enter the estimated cost of each work package. (6) Replace "Category 3 Name" with the name of the third category. In the space marked "Note 6a", enter the sum of the estimated costs of all the work packages listed under the third category. (7) Enter the names of all the work packages to be performed under the third category. Enter the estimated cost of each work package. (8) Replace the "Category 4 Name" with the name of the fourth category. In the space marked "Note 8a", enter the sum of the estimated costs of all the work packages listed under the fourth category. (9) Enter the names of all the work packages to be performed under the fourth category. Enter the estimated cost of each work package. (10) Replace the "Category 5 Name" with the name of the fifth category. In the space marked "Note 10a", enter the sum of the estimated costs of all the work packages listed under the fifth category. (11) Enter the names of all the work packages to be performed under the fifth category. Enter the estimated cost of each work package. (12) Sum the subtotals and indicate this in the space marked "Note 12".

The general contractor will provide the lumber and masonry material and the landscaping, electrical, plumbing, and painting contractors will each provide their own materials. The price quote from these subcontractors will include the cost for material. Subcontractors will estimate their costs using the analogy, parametric, or bottom-up technique. Accuracy will be enhanced because each subcontractor will have many similar projects from which to draw data. Accuracy will be less accurate when subcontractors estimate costs of a project that is significantly different from any they have done in the past. The risk inherent in the process is evident when firm fixed price contracts are established with each subcontractor. In this example, a firm fixed price contract is similar to the departmental budget, except the subcontractor can only invoice the firm a fixed amount regardless of its costs; whereas a department (e.g., engineering) in an organization may charge a project for its costs even when they exceed the department's original estimate (budget).

Table 5.8 shows a functional estimate of the cost to construct a house. The subcontractors provided estimates for each category. Table 5.9 is a template for performing the functional estimate of project cost.

Table 5.7 WBS for the Construction of a House

HOUSE AT 2900 SUNRISE LANE

1.0 General Contractor (Project Management)
   1.1 Purchase building materials
   1.2 Get permits
   1.3 Monitor and control the project
2.0 Land Developer
   2.1 Clear lot of trees and boulders
   2.2 Excavate the hole for basement
   2.3 Complete final grading
   2.4 Landscape the lot
3.0 Masonry Contractor
   3.1 Build basement walls
   3.2 Build fireplace and chimney
   3.3 Pour walkways, driveway, and basement floor
   3.4 Install tile in bathrooms
4.0 Carpentry Contractor
   4.1 Build frame of house and cover with siding/shingles
   4.2 Install interior walls and stairs
   4.3 Install wallboard to walls and ceilings
   4.4 Install hardwood floors
   4.5 Install trim and doors
5.0 Electrical Contractor
   5.1 Install rough wiring
   5.2 Install fixtures and lights
6.0 Plumbing Contractor
   6.1 Install water/sewer lines
   6.2 Install rough plumbing lines
   6.3 Install furnace and air conditioner
   6.4 Install tubs, sinks, and appliances
7.0 Painting Contractor
   7.1 Paint outside trim
   7.2 Paint inside walls and ceilings
   7.3 Paint inside trim

Note: Work packages are listed by function or construction specialty.

Analogy Estimate

This method takes the total cost of a previous similar project and uses it as the baseline from which to estimate the cost of the current project. This method requires that cost data from a previous, similar project must have been retained for each major cost element. Basic to the analogy method is the idea that what we will have

Table 5.8   Example of a Project Budget Determined via the Functional Method[a]

Labor Cost by Function, Department, or Subcontractor		
Landscaping contractor	Note 1	$12,000
Masonry contractor		$18,000
Tile contractor	Note 2	$5,000
Carpentry contractor		$35,000
Plumbing contractor	Note 3	$75,000
Electrical contractor	Note 4	$12,000
Painting contractor	Note 5	$11,000
General contractor		$10,000
	Total Labor Cost	**$178,000**
Materials Cost		
Lumber materials		$45,000
Masonry materials		$13,000
Appliances		$20,000
	Total Materials Cost	**$78,000**
Equipment Cost		
All equipment is subcontractor supplied		
	Total Equipment Cost	**$0**
	Total Project Cost	**$256,000**

[a] *Table 5.9 is the template for this estimate.*

Notes: (1) Shrubs, etc., are supplied by the landscaping subcontractor. (2) Tile and other materials are supplied by the tile subcontractor. (3) Plumbing materials including heating and air conditioning units are supplied by the plumbing subcontractor. (4) Electrical materials are supplied by the electrical subcontractor. (5) Painting materials are supplied by the painting subcontractor.

to pay for something now is probably close to what we paid for it in the past. There are exceptions and caveats to this, and estimating the amount of difference between what we paid in the past and what we will have to pay in the future is the core of this technique.

Example 2: We want to estimate the cost of constructing a one-story ranch house in Alaska. We have the cost data for a house with similar features and size that was built in California last year. Notice that the cost figures for the house in Alaska are differentials (i.e., differences from the cost of the house in California).

Table 5.9 Template for a Functionally Determined Project Cost Estimate

Labor Cost by Function, Department, or Subcontractor	Cost
Note 1	Note 2
Note 1	Note 2
Note 1	Note 2
Note 1	Note 2
Note 1	Note 2
Note 1	Note 2
Note 1	Note 2
	Total Labor Cost Note 3
Materials Cost	
Note 4	Note 5
Note 4	Note 5
Note 4	Note 5
Note 4	Note 5
	Total Materials Cost Note 6
Equipment Cost	
Note 7	Note 8
Note 7	Note 8
Note 7	Note 8
	Total Equipment Cost Note 9
	Total Project Cost Note 10

Instructions: (1) In the Note 1 space, indicate the name of the functions, departments, or contractor performing project work. (2) In the Note 2 space, indicate the cost estimate for each function, department, or contractor. Indicate when the estimate includes the cost of materials needed to perform the work. (3) Sum the costs for labor; indicate this sum as the total labor cost. (4) In the note 4 space, indicate the materials purchased by the project manager. (5) Indicate the cost for each item of material listed under note 4. (6) Sum the costs for materials; indicate this sum as total materials cost. (7) In the note 7 space, indicate the equipment or tools that must be rented or purchased by the project manager. (8) Indicate the cost of equipment or tools listed under note 7. (9) Sum the cost of equipment or tools; indicate this sum as the total equipment cost. (10) Sum the total labor (note 3), total materials cost (note 6), and total equipment cost (note 9). Indicate this sum as the total project cost.

The land for the house in California cost $200,000. We are estimating the cost of the land for the house in Alaska at $50,000; therefore, the differential is –$150,000. Material cost is estimated to be $50,000 cheaper because lumber and masonry material are indigenous to Alaska. However, average labor rates across all the construction labor categories are estimated to be 20% higher; therefore, labor cost is estimated to be $30,000 higher.

California House (Actual)		Alaska House (Estimate)	
Total cost	$500,000	Baseline cost	$500,000
Land	$200,000	Land differential	–$150,000
Material	$150,000	Material differential	–$50,000
Labor	$150,000	Labor differential	+$30,000
		Estimated cost of house in Alaska: $330,000	

The analogy method is a common method of estimating project cost. Its accuracy is good enough to provide an approximation of cost but not nearly accurate enough to serve as the project budget. Notice that example 2 includes just three cost factors. The accuracy of the method is increased as the number of cost factors is increased where each factor accounts for a small portion of the total previous project cost. Table 5.10 shows the analogy method where cost factors are broken into smaller elements. As you might expect, the accuracy is higher when many small cost factors are included. Table 5.11 is a template for the analogy method.

Parametric Estimate

The parametric method uses a single equation to estimate the cost of the entire project. A parametric is sometimes referred to as a cost estimating relationship.

> **Example 3:** The cost to build a one-story wood-frame ranch house with two and a half bathrooms and a one-car garage is estimated at $62.50 per square foot. The equation: cost = $62.50/ft^2. The cost does not include the cost of land. How much do we estimate it will cost to build a one-story ranch that is 1,500 square feet (50 ft long × 30 ft wide)? The parametric cost estimate is $93,750 (1,500 ft^2 × $62.50/ft^2).

Parametrics are used everywhere in cost estimating. Cost analysts using data from previous projects usually determine them. For example, two years ago we built a ranch house that was 60 feet wide by 40 feet long (2,400 ft^2). The total cost excluding the land was $150,000. The cost per square foot was $62.50/ft^2 ($150,000 ÷ 2,400 ft^2). This provides the parametric: cost = $62.50/ft^2. Another ranch house (1,200 square feet) built last year cost $74,000. The cost per square foot for the second and smaller house is $61.67 ($74,000 ÷ 1,200 ft^2). The two estimates are not exactly the same but close enough for parametric estimating. We'll use $62.50/ft^2. From our experience in these two houses, we conclude that the cost

Table 5.10 Example of Project Cost Estimating via the Analogy Method

Previous project: House built in California		Estimate for: House in Alaska	
Total Cost:	**$500,000**	**Cost Baseline:**	**$500,000**
	Cost		**Cost Differentials**
Land	$200,000		−$150,000
Materials			
Blocks & bricks	$20,000		−$3,000
Lumber	$40,000		−$5,000
Insulation	$5,000		+$7,000
Air conditioner	$10,000		−$10,000
Heating unit	$5,000		+$3,000
Plumbing material	$30,000		+$6,000
Appliances	$20,000		+$4,000
Windows & doors	$5,000		+$1,000
Labor			
Laborers	$30,000		+$3,000
Carpenters	$55,000		+$5,500
Masons	$20,000		+$2,000
Electricians	$18,000		+$1,800
Plumbers	$25,000		+$2,500
Painters	$17,000		+$1,400
Equipment	$0		$0
	Estimated Cost of House in Alaska		**$369,200**

Note: Table 5.11 is the template for the analogy estimate.

to build a house as small as 1,200 square feet and as large as 2,400 square feet can be estimated from the parametric: $C = \$62.50/\text{ft}^2$. Please remember that using a single equation to estimate the total cost of a project will provide a very rough approximation of cost.

The accuracy of this technique depends upon how the cost estimating equation was developed and the size of the cost element that it attempts to estimate. Just as with the analogy method, the larger the element of cost being estimated, the greater the error in the estimate. The estimate in Table 5.10 is more accurate than that in example 2 because it breaks costs into smaller pieces. The smaller pieces can be estimated more accurately than can larger elements of cost. The parametric estimate provides a rough approximation of total cost. Actual cost may differ significantly from the single equation estimate. Because of this large potential error, using a single parametric to estimate total project cost should only be used during the initiation phase when such an estimate is useful in the project selection process.

Table 5.11   Template for Estimating Project Cost via the Analogy Method

Previous project: (Note 1)   *Estimate for: (Note 2)*

	Total Cost:	$ (Note 3)	**Cost Baseline:**	$ (Note 3)
Cost Categories (Note 4)		**Cost**	**Cost Differentials**	
Category 1: Materials				
(Note 5)	$ (Note 6)			$ (Note 7)
(Note 5)	$ (Note 6)			$ (Note 7)
(Note 5)	$ (Note 6)			$ (Note 7)
(Note 5)	$ (Note 6)			$ (Note 7)
(Note 5)	$ (Note 6)			$ (Note 7)
(Note 5)	$ (Note 6)			$ (Note 7)
Category 2: Labor				
(Note 8)	$ (Note 9)			$ (Note 10)
(Note 8)	$ (Note 9)			$ (Note 10)
(Note 8)	$ (Note 9)			$ (Note 10)
(Note 8)	$ (Note 9)			$ (Note 10)
(Note 8)	$ (Note 9)			$ (Note 10)
(Note 8)	$ (Note 9)			$ (Note 10)
Category 3: Equipment				
(Note 11)	$ (Note 12)			$ (Note 13)
(Note 11)	$ (Note 12)			$ (Note 13)
	Estimated Cost of (Note 2)			$ (Note 14)

Instructions: (1) Indicate the name of the project which costs are being used as the baseline for the analogy estimate. (2) Indicate the name of the future project which costs are being estimated. Do this in two places. (3) Indicate the total cost of the previous project in two places. (4) Indicate the major categories of costs. The format includes the materials, labor, and equipment categories. Change the names as appropriate. Add more categories if necessary. (5) List the elements of cost under the first category. (6) Indicate the cost of each element in the first category from the previous project data. (7) Indicate the cost differential for each element of cost. For example: the cost of lumber in the previous project was $10,000. We estimate that the cost of lumber for the future house will be $1,000 more. Enter +$1,000 in the cost differential column. Another example: cost of masonry materials in the previous house was $20,000, and we estimate that the cost will be $2,000 less in the future house. Enter −$2,000 into the cost differential column. Cost differential means difference in cost from the previous project; a plus (+) means it will cost more and a minus (−) means it will cost less than the previous project. (8) List the elements of cost under the second category. (9) Indicate the cost of each element in the second category from the previous project data. (10) Indicate the cost differential for each element of cost in category 2. (11) List the elements of cost under the third category. (12) Indicate the cost of each element in the third category from the previous project data. (13) Indicate the cost differential for each element of cost in category 3. (14) Sum the cost baseline and all the differentials to determine the estimated cost of the future project. Enter this number in the Note 14 space.

Note: Refer to the example in Table 5.10.

Scalability and Extrapolation

The usefulness of a parametric depends upon the range of data from which it was developed. Using the cost estimating equation outside the range of data from which it was determined is called extrapolation. The parametric for the cost of a ranch house as $62.50 per square foot was developed from the experience of building two previous houses. The smaller house was 1,200 square feet and the larger was 2,400 square feet; 1,200 and 2,400 square feet are the smallest and largest house sizes for which this parametric applies. Estimating the cost of a house with a size between 1,200 and 2,400 square feet seems justified. Scalable refers to the idea that the equation for the line is a straight line; this is also called linear relationship. A linear relationship means that we assume that the costs are proportional to house size. If we increase the house size, for example, by 100%, the cost increases by 100%. The assumption of linearity is justified over the range 1,200 to 2,400 square feet because we have two data points demonstrating this that define the cost estimating line. If the parametric were based on the cost of only one house, e.g., 1,200 square feet, how would we know that the cost of $62.50 per square foot would be applicable to a house that is 2,400 square feet? Using a parametric that is based on one point of past information is risky if the size of the house being estimated is significantly different (larger or smaller) than the size of the house from which the single-point parametric was developed.

Extrapolation means going beyond the data. The largest house is 2,400 square feet. Will the relationship (cost = $62.50 per square foot) provide a reasonable estimate for a house that is 7,000 square feet? A house of this size is a mansion. Mansions include many features not found in a 1,200 or 2,400-square-foot house, such as a swimming pool, a library, an entertainment center, a gourmet kitchen, etc. These features will significantly increase the cost per square foot. Extrapolating a little beyond the data range of the parametric is justified. How far you can extrapolate beyond the data is a matter for conjecture best left to professional cost analysts.

Notes

1. "Adding Risk into Project Estimates — PERT vs. Monte Carlo," by Guy L. De Furia, Ph.D., PMP. (Article first appeared in the April 2007 issue of *ESI Horizons.* © ESI International, 901 North Glebe Road, Suite 200, Arlington, VA 22203; www.esi-intl.com. With permission.)
2. For a discussion of how PERT can be used to estimate schedule and cost risk, see the article cited in note 1.
3. Formulas 5.3 and 5.4 give the instructions for determining effort time at the 95% level of confidence. (Eq. 5.3): $T_e + 2\sigma$ estimates the maximum effort time at 95% confidence. (Eq. 5.4): $T_e - 2\sigma$ estimates the minimum effort time at 95% confidence. When we are only interested in the maximum effort time (but not the minimum), we use Eq. 5.3; in this circumstance, the level of confidence becomes 97.5%.

Chapter 6

Project Schedule

Mega Recipes for Developing the Schedule

Activity 1: Get viable estimates of work package duration time for every work package listed on the work breakdown structure (WBS).

Activity 2: Develop the network (precedent) diagram.

Activity 3: Draw the Gantt chart. Develop a resource Gantt if necessary to highlight the people resources needed to maintain the schedule.

Activity 4: Cross-check the schedule against the resource plan.

Activity 5: Get the schedule (network diagram and Gantt chart) approved.

Purposes of the Scheduling Activities

The purposes of the scheduling activities are to develop a realistic schedule for accomplishing the project and to assure that the significant stakeholders understand and accept the resource commitments associated with maintaining the schedule.

Role of the Project Manager

1. Get realistic estimates of work package duration times.
2. Lead the team in its efforts to develop a viable schedule.
3. Facilitate upper management's understanding and acceptance of the schedule and the resource commitments associated with it.

Outputs of Schedule Planning

The desired outcomes of the schedule planning activities are a viable schedule for the conduct of the project represented by the network diagram and Gantt charts plus upper management's understanding, acceptance, and commitment to the schedule.

Schedule-Planning Problems

1. The biggest problem during the scheduling planning process is one of human inertia — the "we don't do it that way here!" syndrome. The biggest threat to developing a viable schedule is the belief that it is not necessary to include risk in the schedule or that it is too much work to do so. A study by the author[1] concluded that the confidence one should have in project durations based on single point estimates for each work package is less than 50%. It concluded that project schedules based on the PERT (Program Evaluation Review Technique, described later in this chapter) average time to complete each work package should have confidence levels of about 50%. The ability to increase the confidence levels in total project duration times to the 95% level (or any other desired level) is easy to achieve if the PERT three-point estimates of duration times are used. From the author's experience, it appears that few organizations use the PERT or Monte Carlo methods to include risk in the determination of the project duration time. To increase the confidence in the project duration time, use the PERT or Monte Carlo techniques. (PERT is described later in this chapter.)

2. One of the desired outcomes of the schedule-planning activities is upper management understanding and commitment to providing the resources required to sustain the schedule. A potential problem: managers or sponsors do not see the numbers of people with specific skill sets required during specific periods of time, and fail to make arrangements to support these requirements. To address this potential problem, the project manager should have a resource Gantt prepared to make this clear. Design and facilitate the briefing to assure that the resource implications of the schedule are clear. Get the schedule and resource requirements approved.

Activity 1: Get Viable Estimates of Work Package Duration

This topic is broken into two sections: definitions and basic relationships among the definitions, and how to estimate work package duration time.

Definitions and Basic Relationships

Time is one of the basic dimensions of every project. Before you start to develop the project schedule you must get accurate estimates of the time required to accomplish each work package listed on the work breakdown structure. Most of the estimates of work package duration time are going to be estimates that others give you. To get useable estimates, the project manager and the people providing the estimates must understand the definitions of different kinds of time.

Normal work days (sometimes called, **working time or working calendar**) refer to the calendar of days that the organization is open for work and the number of hours per day that employees work. Many organizations work five days per week — Monday through Friday and not Saturday and Sunday. Most organizations do not work on national holidays such as New Year's Day. Some organizations require employees to work eight hours per day. The normal work days must be defined before you establish when each work package will be performed because you should not schedule work when the organization is not open for work. Normal work days may vary among occupational specialties within the same organization. Production and administrative workers usually work five eight-hour days per week while the security function may be performed seven days per week. This means that the project may schedule administrative workers to perform work packages only on Monday through Friday. Work on Saturday and Sunday is scheduled only under extraordinary conditions.

Effort time is the time estimated by an experienced worker to complete a work package. The single greatest reason for projects completing behind schedule is poor estimates of the time required to complete the work packages. To ameliorate this, get estimates from people who have experience in performing the work required by the work package. Only accept time estimates from people who are experienced in the kind of work required by the work package — think of these people as 100% proficient in the work required by the work package.

Actual time is the real time that will be required to perform the work package. The actual time will differ significantly from the estimate of effort time if the person who performs the work package differs significantly in experience from the person around whom the effort time was originally estimated.

> **Example 1:** The work package requires the skills of a person who can write software code in a particular language. The estimate of effort time from an experienced code writer is 40 hours. For purposes of our example, the estimate is based on a 100% proficient, experienced or productive worker — the person who gave the estimate. If a person less proficient in code writing (e.g., 50% proficient) does the work package, it will greatly impact the actual time taken to achieve the result. The

relationship is expressed as follows:

$$\text{Actual time} = \text{effort time} \div \text{proficiency} \qquad (6.1)$$

$$\text{Actual time} = \text{effort time} \qquad (6.2)$$

Example 1 continued:

From Equation 6.2: 40 hours ÷ 1.00 = 40 hours actual time if the experienced estimator performs the work package

From Equation 6.1: 40 hours ÷ .50 = 80 hours actual time if the 50% proficient worker performs the work package

Changing the person who performs the work package from the person around whom the effort time estimate is based caused significant underestimation of actual time. The project schedule is based on accomplishing the work package in one week (40 hours) but the work package actually required two weeks because a less experienced, less proficient worker was used. The project has lost a week of time because of this.

When you get an estimate of effort time to perform a work package, negotiate the decision about who will actually perform the work package. (Record this in the resource plan.) The person who gives the estimate of effort time should be the same person who performs the work package. If another person performs the work package, be sure that this person has the same level of proficiency as the person who made the estimate of effort time, i.e., can perform the work package in the estimated effort time. The functional manager or supervisor may not be aware of the dynamics at work here, so take the time to explain why it is important to "lock in" the person who will perform the work package.

$$\text{Labor cost} = \text{actual hours} \times \text{loaded labor rate} \qquad (6.3)$$

Example 2: The effort time for a work package is estimated at 40 hours. The loaded labor rate (cost per hour of labor) is $50 per hour. The budgeted cost of the work package from Equation 6.3 is $2,000 (40 hours × $50 per hour). However, a less proficient person actually performed the work package and needed 80 hours to complete it. The loaded labor rate of this less proficient person is $40 per hour; the actual cost for this work package from Equation 6.3 is $3,200 (80 hours × $40 per hour). In this example, using a less proficient person to perform the work package resulted in it taking twice as long as planned and costing $1,200 more than budgeted. The project will be charged for actual time, not the estimated effort time. It will be charged the actual cost, not the budgeted cost.

Duration time is often used interchangeably with effort time. In some cases, effort and duration time are the same; but in many cases they are not. Duration time is defined as the estimated time from the start to the end of a work package but not including non-work days. The equation is

$$\text{Duration time} = (\text{effort time} \div \text{proficiency}) \div \text{availability} \qquad (6.4)$$

$$\text{Duration time} = \text{actual time} \div \text{availability} \qquad (6.5)$$

The effect of availability upon duration time is shown in example 3.

Example 3: Effort time for a work package is estimated at 40 hours. It is assumed that this will be a full-time effort. The work package is completed in 40 hours but it is done on a half-time (.5) basis because the worker must work on other projects also. The schedule calls for completing this work in one week. The budget for the work package is $2000 (40 hours of effort × $50 per hour). Equation 6.5 shows us that the duration time is 80 hours (40 hours ÷ .5). The project will be charged for 40 hours of work spread over two weeks. Availability affects the schedule but not the cost. In this example, the work package cost was $2000 as originally estimated but it took two weeks instead of one to complete the work package.

Example 4: This example shows what happens when both work proficiency and availability change. Effort time is originally estimated at 40 hours for a fully competent worker (100% proficiency). The loaded labor rate for this fully competent worker is $50 per hour. A full-time effort is assumed. In actuality, a less proficient worker performs the work package (e.g., proficiency = .5) and the work is performed on a half-time basis (e.g., availability = .5); the loaded labor ratio for this second person is $40 per hour. The interactive effect of both reduced proficiency and less than a full-time effort is

$$\text{Actual hours} = \text{effort time} \div \text{proficiency}$$

$$\text{Actual hours} = 40 \text{ hours} \div .5 = 80 \text{ hours}$$

$$\text{Labor cost} = \text{actual hours} \times \text{loaded labor rate}$$

$$\text{Labor cost} = 80 \text{ hours} \times \$40 \text{ per hour} = \$3200$$

$$\text{Duration time} = \text{actual hours} \div \text{availability}$$

$$\text{Duration time} = 80 \text{ hours} \div .5 = 160 \text{ hours}$$

Instead of being completed in 40 hours (one week) this work package will require 160 hours (four weeks) of half-time effort, although the worker will charge for 80 hours of actual time. Because the assumptions of full proficiency and full-time effort did not apply, the effort came in three weeks late and $1200 over budget.

Duration time equals effort time when proficiency and availability are 100% (i.e., 1.0). The estimates of duration time and effort time will differ significantly when the assumptions of 100% proficiency and a full-time effort (100% availability) are not met. The project schedule and budget will be significantly violated when proficiency and availability are not explicitly negotiated at the time the schedule and budget are being established.

Examples 1 through 4 demonstrate why it is important that both the project manager and functional manager have a clear and mutual understanding of these time definitions. It is imperative that the negotiation between the project manager and the functional manager be clear about the proficiency of the worker around whom the effort time is estimated, the proficiency of the person who will actually complete the work package, and the issue of availability.

The project manager's task is not complete until three pieces of information are determined for every work package:

1. Effort time estimate from an experienced worker
2. Who will perform the work package is established
3. The productivity and availability of the performer are established

This information must be added to the resource plan discussed in Chapter 7.

How to Estimate Work Package Duration Times

Single-point estimates were used in examples 1 through 4. The methodology developed by the U.S. Navy provides a desirable alternative to the single-point estimate.

Project managers know that completing a project on time includes identifying and managing the risks associated with the work packages on the critical path. The process for doing this includes:

1. Estimating the duration time for every work package on the work breakdown structure.
2. Drawing the network diagram based on your knowledge of chronological relationships among the work packages.
3. Performing the forward and backward passes to identify the critical path. (How to do the forward and backward passes is covered later in this chapter.)

The completion of the forward pass determines the project duration time. The completion of the backward pass provides the information to identify those work packages on the critical path. This procedure is straightforward and simple.

However, this simple procedure has significant limitations. Often, there is confusion about how to interpret the effort time estimates. When we ask the work package manager to estimate how long it will take to complete a work package, we don't know if the answer we get is the average, the shortest, or the longest time to complete the work package. Some people will provide the average time to complete the work package because this seems reasonable; some will give the longest time to be on the safe side. Some people (admittedly, only a few) will give the shortest time because they are incurable optimists. Often, the estimates include a little "fudge factor" just to be safe. Unfortunately, receivers of the estimates have no way of judging the confidence that is appropriate to the single estimate of work package duration. What is needed is a way to numerically estimate our confidence in each estimate of time and a way of translating our confidence or lack of confidence into the number of days the project may be delayed.

The U.S. Navy recognized this dilemma in the 1950s when it was developing its scheduling tools for use in the building of the first Polaris submarine.[2] The Navy recognized that asking for three estimates of duration time for each work package would provide the information with which to:

1. Develop a useable estimate of work package mean duration time
2. Estimate the confidence associated with this mean duration time
3. Estimate the work package duration time at varying levels of confidence

This last capability allows us to add risk into our estimates of duration time. Later in the chapter we will see how this may be done at the total project duration level.

PERT: Program Evaluation Review Technique is the name the Navy gave to the collection of equations that address the dilemma above:

$$T_e = (P + 4[ML] + Op) \div 6 \qquad (5.1)$$

$$\sigma = (P - Op) \div 6 \qquad (5.2)$$

where T_e is the average effort time estimate, P is the pessimistic estimate of work package effort time, ML is the most likely estimate of effort time, and Op is the optimistic estimate. The Greek sigma (σ) is the symbol for standard deviation.

> **Example 5:** The following three estimates were received from an experienced worker: pessimistic time to complete the work package is seven days (56 hours), most likely estimate is 40 hours, and the optimistic estimate is 32 hours.
>
> Average effort time $T_e = (P + 4[ML] + Op) \div 6$
>
> $T_e = (56 + 4[40] + 32) \div 6 = 41.3$ hours
>
> Standard deviation $\sigma = (P - Op) \div 6$
>
> $\sigma = (56 - 32) \div 6 = 4$ hours

The standard deviation, 4 hours in example 5, is important in two ways: (1) it provides a way of judging the confidence we should have in its mean (41.3 hours in our example), and (2) it provides a means of adding risk to our estimate of work package effort time. When the standard deviation approaches 30% of its mean, we should be alerted to the strong possibility that the work package will be completed in a time significantly different from the mean — either very late or very early. In examples 5, the mean is 41.3 and the standard deviation is 4. Four is 9.68% of the mean (4 ÷ 41.3), so we can be confident in using the mean of 41.3 hours.

We can be 50% confident that the work package will be completed in 41.3 hours; 50% is the confidence associated with completing any work package in its average duration time. We may use the standard deviation to provide an estimated effort time for which we have a higher level of confidence.[3] The equations for maximum and minimum effort times at 95% confidence and 90% confidence are as follows:

$$\text{Maximum effort time}_{95} = T_e + 2\sigma \tag{6.6}$$

$$\text{Minimum effort time}_{95} = T_e - 2\sigma \tag{6.7}$$

$$\text{Maximum effort time}_{90} = T_e + 1.65\sigma \tag{6.8}$$

$$\text{Minimum effort time}_{90} = T_e - 1.65\sigma \tag{6.9}$$

For example 5, we can easily increase our confidence in completing the work package on time by using Equations 6.10 and 6.11 where $T_e = 41.3$ hours and $\sigma = 4$ hours:

$$\text{Maximum effort time at 90\% confidence} = T_e + 1.65\sigma = 41.3 + 1.65(4) = 47.9 \text{ hours}$$

$$\text{Minimum effort time at 90\% confidence} = T_e - 1.65\sigma = 41.3 - 1.65(4) = 34.7 \text{ hours}$$

Instead of using an effort time of 41.3 hours in our schedule, of which we are only 50% confident, we can use 47.9 hours, an effort time in which we are 90% confident.

Success as a project manager means developing a viable, achievable schedule and this requires good estimates. One way to encourage conscientious estimates is to use an estimating worksheet similar to the ones in Tables 5.3 and 5.4.

Fortunately, spreadsheet applications can be used to perform the PERT calculations. Figure 5.1 is an Excel® template for estimating effort times and costs of work packages at 50%, 90%, and 95% levels of confidence.

The approach described previously where we add one or more standard deviations to the average effort time to complete a work package is the technique that increases our confidence that we can complete the work package in its estimated effort time. There is an alternative approach whereby we increase the critical path duration to increase our confidence in completing the project on time. This method is discussed in Chapter 9 (Figure 9.6).

Activity 2: Develop the Network (Precedent) Diagram

The time dimension for a project is defined by three documents, listed in order of development:

1. The network diagram
2. The Gantt chart (described in activity 3)
3. The milestone chart (described in activity 3)

The network diagram is a picture showing the chronological relationships among the work packages. It shows which work packages can run concurrently with another work package and which work packages must be performed after other earlier work packages are completed. The relationship among work packages is called the predecessor and successor relationships (precedent relationship for short). "Predecessor" refers to a work package that must be performed before another work package can be started. "Successor" work package refers to a work package that must be started after a previous one is completed. "Network diagram" is a generic name; the names "precedent diagram" and "activity on the node (AON) diagram" are also names used for this picture.

Rules for Drawing a Network Diagram

1. Every work package shown on the WBS must be shown on the network diagram and the Gantt chart.
2. You need three kinds of information before you can draw the network diagram:
   a. A list of all work packages titles (shown on the WBS)
   b. Estimated duration time for every work package
   c. The precedent relationships among the work packages.
   The knowledge of the precedent relationships among the work packages comes from the team's experience. If you have ever been involved in building a house, you know that the hole for the basement must be excavated before the basement walls can be constructed and the wood frame of the house cannot be constructed until the basement walls are completed.
3. Each work package is represented by a rectangle. Arrows connect the rectangles. The tail of the arrow touches the predecessor work package and the head of the arrow touches the successor work package.
4. Every work package must have at least one arrow into the work package and at least one arrow coming out of the work package. The only exceptions are the rectangles titled "start" and "finish." These are not really work packages; they are milestones. The "start" rectangle has one or more arrows coming out of the right edge but none going into it. The "finish" rectangle has one or more arrows going into the left edge but none coming out of it.

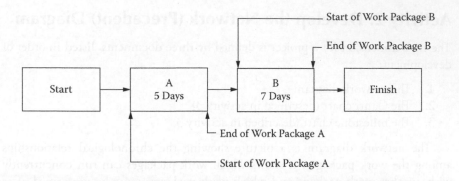

Figure 6.1 Network diagram showing predecessor and successor work packages. Work package A is the predecessor to work package B. Work package B is the successor to work package A. The left edge of the work package rectangle is the start side and the right edge is the finish side. The arrow point always touches the successor and the tail of the arrow always touches the predecessor.

5. Include the work package identifier (number or letter) and the estimated duration inside each rectangle. The work package identifier is included in the rectangle because the title often will not fit in the rectangle. Figure 6.1 shows the basic notations to the network diagram.

6. Time is on the horizontal axis. Later-occurring work packages should be further to the right than those occurring earlier. (This is sometimes violated to achieve layout efficiency.)

7. The arrow connecting two work packages shows the precedent relationship between them. The point of the arrow always touches the successor work package. The tail of the arrow touches the predecessor work package. This rule must not be violated. Computer-generated network diagrams sometimes violate this rule by showing a picture of one kind of relationship between two work packages but labeling the relationship as a different relationship. When this occurs, interpret the relationship as indicated by the label. Avoid, if possible, using relationship labels to describe precedent relationship because many people find them confusing. See Figure 6.3[B] for an example.

Precedent Relationships among Work Packages

Three kinds of precedent relationships may exist between two adjacent work packages:

1. **Finish-to-start (FS):** The finish-to-start is the most common relationship but you must use the relationship required by the logic connecting the two adjacent work packages. For example, work package 4.8, "Shingle the Roof," must follow work package 4.1, "Build House Frame." Therefore the

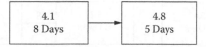

Figure 6.2 Finish-to-start relationship. In the FS notation, the F (the first letter) refers to the predecessor work package and the S (the second letter) refers to the successor work package. The FS notation means that the arrow must start at the finish side of work package 4.1 and go directly to the start side of work package 4.8. Therefore, the arrow is drawn with its tail (where the feathers would be if it were a real arrow) touching the finish side of 4.1 and the point of the arrow touching the start side of 4.8. The arrow point tells us that work package 4.8 is the successor and 4.1 is the predecessor.

relationship between work package 4.8 and work package 4.1 must be a finish-to-start with 4.8 the successor. See Figure 6.2.

2. **Finish-to-finish (FF):** Sometimes two work packages must finish on the same day. For example, we have a construction project in a remote area. "Deliver Materials" is work package C and "Deliver Workers" is work package D. These two work packages will take different amounts of time to complete but both must be completed on the same day. A finish-to-finish relationship would seem appropriate between work packages C and D. See Figure 6.3.

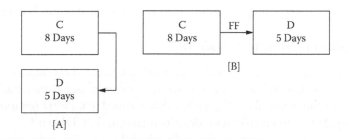

Figure 6.3 Finish-to-finish relationship. Diagram [A] is the usual way to represent the finish-to-finish relationship. In the FF notation, the first F refers to the predecessor work package and the second F refers to the successor. Again the arrow point between the work packages tells us that D is intended to be the successor and C the predecessor because the arrow point always touches the successor. In the absence of a number on the arrow, D and C must end on the same day. Diagram [B] shows the way a computer-generated diagram would represent the FF relationship. Notice that the arrow shows a finish-to-start relationship but the FF label on the arrow tells us that the relationship is really a finish-to-finish relationship. The label overrides the picture. The label FF refers to the finish-to-finish relationship; it does not refer to free float, which is never labeled as FF on the network diagram.

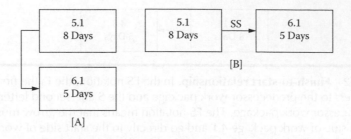

Figure 6.4   Start-to-start relationship. Diagram [A] is the usual way to represent the start-to-start relationship. In the SS notation, the first S refers to the predecessor work package and the second S refers to the successor. Again the arrow point between the work packages tells us that work package 6.1 is the successor and 5.1 is the predecessor. In the absence of a number on the arrow, work packages 5.1 and 6.1 should start on the same day. Diagram [B] shows the way a computer-generated diagram would represent the SS relationship. Notice that the arrow shows a finish-to-start relationship but the SS label indicates that the relationship is really a start-to-start. The label overrides the picture.

3.   **Start-to-start (SS):** Sometimes two work packages can start on the same day. Such is the case for work package 5.1 "Install Rough Wiring" and 6.1 "Install Rough Plumbing." A start-to-start relationship would seem appropriate between work package 5.1 and work package 6.1 because the schedule planners want them both to start on the same day. See Figure 6.4.

Day Number Notation System

There are four dates associated with each work package. These appear in the four corners of the rectangle. There can only be one number in each corner. Each number represents an important date. Figure 6.5 shows this day number notation system. Converting day numbers to dates is described later and in Table 6.1.

The procedure to determine the early schedule (i.e., the early start [ES] and early finish [EF] days) is called the forward pass, like the forward pass in football. The procedure to determine the late schedule (i.e., the late start [LS] and late finish [LF] days) is called the backward pass.

Rules for the Forward Pass

1.   **Always add,** which means to get from one rectangle to the next you add the early finish number from the predecessor to the number on the arrow to get the number you place in the early start corner of the successor. If there is no number on the arrow, read it as a zero. Always commence the forward pass at the start box and proceed to the right. In Figure 6.6, 0 (start box) + 0

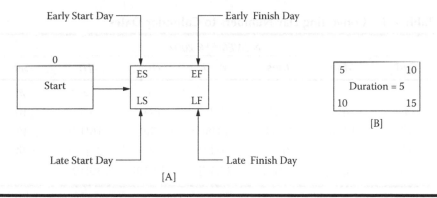

Figure 6.5 Day number notation system. In diagram [A], the rectangle for every work package on the network diagram includes the following day numbers: the early start (ES), early finish (EF), late start (LS), and the late finish (LF). The forward pass procedure determines the early start and early finish day numbers. The backward pass procedure determines the late start and late finish day numbers. The most common convention is to initiate the forward pass with the number 0 above the start box. With this convention the work package in diagram [B] is interpreted as follows: the early start will be on the morning of the day after the number in the early start corner; in this case the work package will actually start on the morning of day 6. The early finish will be on the evening of day 10. This applies to all the work packages in the network diagram. The actual early start or late start day is always the morning of the next work day. The actual early finish or late finish day is always the evening of the day numbered in the upper right or lower right corner, respectively. Table 6.1 shows how to convert day numbers to calendar dates.

(on the arrow) = 0 (ES of J); put this number is the early start corner of work package J. To determine the early finish number, add the duration to the early start number. In work package J, 0 (ES of J) + 10 (the duration) = 10 (the EF of J). Repeat this procedure of going from one rectangle to the next and from the early start corner of one work package to the early finish corner of the same work package. Continue this until you arrive at the finish box: 10 (EF of J) + 0 = 10 (ES of K); 10 (ES of K) + 5 (duration of K) = 15 (EF of K). The finish day of the project: 15 (EF of K) + 0 = 15; 8 (EF of L) + 0 (on arrow) = 8 (finish box).

2. **Put the numbers in the top corners of the rectangles.** The number in the upper left is the early start and the upper right is the early finish. Note that the numbers in Figure 6.6 are in the upper corners or just outside the box — but clearly located near the early start and early finish corners.

3. **Follow the direction of the arrows,** which means that the number determined as the early finish from the predecessor work package must be

Table 6.1   Converting Day Numbers to Calendar Dates

			NOVEMBER 2007			
Sun	Mon	Tues	Wed	Thur	Fri	Sat
				1	2	3̸
4̸	5[1]	6[2]	7[3]	8[4]	9[5]	1̸0̸
1̸1̸	12[6]	13[7]	14[8]	15[9]	16[10]	1̸7̸
1̸8̸	19[11]	20[12]	21[13]	2̸2̸	23[14]	2̸4̸
2̸5̸	26[15]	27[16]	28[17]	29[18]	30[19]	

Note: We want to convert the day numbers in figure 6.14 to calendar dates. Non-work days are indicated on the calendar above by a double strikethrough. We want to start the project on Monday, November 5th. In figure 6.14, work package A has day numbers ES 0 and EF 5. This means that it starts on day 1, which is Nov. 5th. Counting with day 1 on Nov. 5, day 2 on Nov. 6, day 3 on Nov. 7, day 4 on Nov. 8, and ending on day 5, Nov. 9. (Refer to figure 6.5.) To help speed up the conversion process and reduce errors, the day numbers are shown on the calendar above. They are indicated in the brackets [ ]. Work package B starts on Nov. 1 (day 1) and ends on Nov. 16 (day 10). Work package C starts on Nov. 12 (day 6) and ends on Nov. 20 (day 12). Work package D starts on Nov. 6 (day 2) and ends on Nov. 15 (day 9). Work package E starts on Nov. 19 (day 11) and ends on Nov. 30 (day 19). Work package F starts on Nov. 12 (day 6) and ends on Nov. 19 (day 11).

placed where the arrow indicates. In Figure 6.6, the arrow from work package J to work package K goes from the finish side of J to the start side of K. This means that the early finish number from J must be modified by the number on the arrow and proceed to the start side (that is the early start corner) of K. Note in Figure 6.12 that the early finish of work package C (12) goes directly to the early finish of F because the arrow goes to the finish side of F.

4. **When you come to a path convergence on the forward pass, you determine all the possible number choices and select the higher or highest number.** A path convergence is where two or more arrow points come together. In Figure 6.6, there is a path convergence at the finish box. There is the arrow coming from work package L and another coming from work package K. Because there are two arrows, we must find two numbers to place over the finish box and select the higher number. From L, 8 (EF of L) + 0 (on the arrow) = 8 (finish number). From K, 15 (EF of K) + 0 (on the arrow) = 15 (finish number). Because 15 is larger than 8, we keep 15 and cross off or discard the 8. The number you get at the end of the forward pass (number over the finish box) is the project duration.

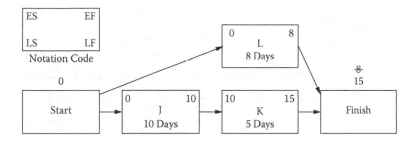

Figure 6.6 Forward pass with path convergence. Rules for the forward pass: (1) Always add; commence over the start box with the number zero. To get from one box to the next, add the number on the arrow. If there is no number on the arrow, read this as a zero. To get from one corner (ES) to the other corner (EF), add the duration. (2) Show the numbers on the tops (inside or outside) of the rectangle. These numbers are referred to as the early schedule. (3) Follow the direction of the arrows; that is, put the determined number on the side that the arrow point is touching. The arrow coming out of work package J goes to the start side (left side is the start side) of work package K, therefore the 10 (10 EF of J + 0 = 10 ES of K) must go in the start corner (upper left). (4) At a convergence, determine one number for each arrow and keep the larger or largest. There is a convergence at the finish box. The arrow from L provides an 8 (8 EF of L + 0 on the arrow = 8 finish box) and the arrow from K provides the number 15 (15 EF of K + 0 on the arrow = 15 finish box). The 15 was retained and the number 8 was discarded. (The 8 is shown with a double strikethrough.)

Rules for the Backward Pass

1. **Always subtract.** Start the backward pass at the finish box and proceed to the left. To get from one box to the next, always subtract the number on the arrow. If there is no number on the arrow, read it as a 0. In Figure 6.7, 15 (finish) – 0 (on the arrow) = 15 (LF for work package L and work package K; 15 (LF of L) – 8 (duration of L) = 7 (LS of L); 15 (LF of K) – 5 (duration of K) = 10 (LS of K). Continue this procedure until you arrive at the start box: 10 (LS of K) – 0 (on the arrow) = 10 (LF of J); 10 (LF of J) – 10 (duration of J) = 0 (late start of J); 0 (LS of J) – 0 (on the arrow) = 0 (start box); 7 (LS of L) – 0 (on arrow) = 7 (start box). Notice there are two arrows backing into the start box, therefore we found two numbers: 7 and 0. Rule 4 directs us to keep the 0.

2. **Put the numbers in the lower corners.** Rule 2 requires that numbers determined during the backward pass procedure are always put in the lower corners. In Figure 6.7, the numbers in the upper corners were omitted to reinforce the idea that the backward pass is concerned only with the numbers in the lower corners. In a real diagram, the upper numbers would be shown.

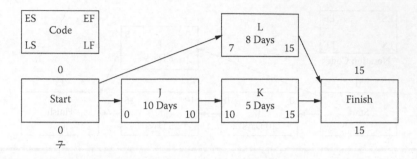

Figure 6.7 Backward pass with path convergence. The early schedule numbers (ES and EF) shown in Figure 6.1 have been omitted from this figure. Rules for the backward pass: (1) Always subtract; commence under the finish box with the number from on top of the box: 15. Proceed to the left. To get from one box to the next, subtract the number on the arrow. If there is no number on the arrow, read this as a zero. To get from one corner (LF) to the other corner (LS), subtract the duration. (2) Show the numbers in the bottom corners (inside or outside) of the rectangle. These numbers are referred to as the late schedule. (3) Proceed in the direction opposite to that of the arrows; that is, put the determined number on the side that the back of the arrow (where the feathers would be if it were a real arrow) is touching. The arrow backing out of work package K goes to the finish side (right side is the finish side) of work package J, therefore the 10 (10 LS of K – 0 on the arrow = 10 LF of J) must go in the late finish corner (lower right). (4) At a convergence, determine one number for each arrow and keep the smaller or smallest. There is a convergence at the start box. The arrow from L provides a 7 (7 LS of L – 0 on the arrow = 7 finish box) and the arrow from J provides the number 0 (0 LS of J – 0 on the arrow = 0 finish box). The 0 was retained and the number 7 was discarded. (The 7 is shown with a double strikethrough.)

3. **Always proceed in a direction opposite to the arrow-like swimming upstream.** Following this carefully will help you put the numbers in the correct corners. For example, in Figure 6.7, the finish box has the number 15. This number has to go in two directions because there are two arrows. One arrow backs into the finish side of K; this is why there is a 15 in the late finish corner. The 15 couldn't be put in the early finish corner because this corner is on the top, and rule 2 directs that numbers during the backward pass are always placed in the lower corners. Likewise, the 7 in the late start corner of work package L must go to the bottom of the start box. This rule becomes very important in a more complicated diagram like the one in Figure 6.13. The 19 (late finish) in work package F must "travel" against the direction of the arrow to get to the late finish of work package C; 19 (late finish of F) minus — 1 (on arrow) = 20 (late finish of C). We

know that the 20 goes in the finish corner because the arrow backs into the finish side of C. Can't put the number on the top corner (replacing the 12) because all numbers in the backward pass must go in the lower corners; this is rule 2.

4. **At a path convergence, you must find the number associated with each arrow and keep the lower or lowest number.** In Figure 6.7, there is a path convergence on the backward pass at the start box. There is an arrow from work package L and another from work package J. Work package L produces a 7 for the start box and work package J produces a 0. Keep the 0 and reject the 7; this is in accordance with rule 4.

Tips to Remember: The forward pass is like a walk through the forest; you must travel along each path and you can't jump ahead. In Figure 6.6, this means you go from start to early start of L to early finish of L to finish. Likewise, you go from start to early start of J to early finish of J to early start of K to early finish of K to finish. You can't skip ahead. It's the same for the backward pass.

The number over the finish box is the project duration.

You can't start the backward pass unless the forward pass is completed first. A complete forward pass means there is one number and only one number in each upper corner.

You have made no errors if the number at the start box resulting from the backward pass is the same number as the number with which you started the forward pass. They should both be zero.

The kinds of errors that are made in network diagramming are simple and human: simple errors of addition or subtraction and forgetting to notice and follow the rules at a convergence.

Schedule Adjusters

Lag and lead are two ways in which the timing of work may be modified to meet special requirements.

Lag is a positive number on an arrow going into a work package that delays the start of the successor. Sometimes the nature of the work requires a delay between the completion of the predecessor work package and the start of the successor work package. For example, we pour a concrete floor in a factory building on day 10. This is work package D. The next work package requires that we bore holes in the concrete. This is work package E. However, we are told by the civil engineer that the boring cannot proceed until the concrete has cured for two days. Work package E must wait two days after D is completed before it can start. To specify this two-day wait, we place a number two on the arrow connecting work packages D and E. This situation is called lag. Adding a lag on the arrow between two work packages

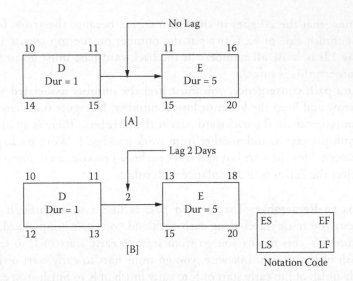

Figure 6.8   Network diagram with lag. A lag is a positive number on the arrow connecting two work packages, thereby delaying the start of the successor work package by an amount of time equal to the lag number. In diagram [A], there is no lag; the results of the forward and backward passes are shown. In diagram [B], there is two days of lag. Note that work package E has been delayed from day 11 to day 13. Note that 2 was added in the forward pass (11 EF of D + 2 = 13 ES of E) and subtracted in the backward pass (15 LS of E – 2 = 13 LF of D). A lag is always a positive number. Lag can be specified in any of the three precedent relationships: finish to start, start to start or finish to finish. The rules for the forward and backward passes always apply.

will delay the start of the successor by the amount of the lag. Refer to Figure 6.8 for a demonstration of lag.

Lead is a negative number on the arrow going into a work package that we want to start before the completion of the predecessor. Sometimes the work requires or the team wants to start the successor work package before the predecessor work package is completed. For example, the schedule is very tight; work package 6.5 calls for doing five days of data gathering. Its successor 6.6 calls for 10 days of doing analysis of the data gathered. One of the team members notes that we could start to do the analysis (work package 6.6) two days before work package 6.5 is complete because three days of data gathering will produce enough information with which to start the analysis. To start the successor work package before the predecessor is complete requires a lead. Figure 6.9 shows the effect of adding a two-day lead into the schedule.

Figure 6.10 demonstrates the forward pass for a network that includes lag and lead plus path convergence. Make sure you understand how the early schedule

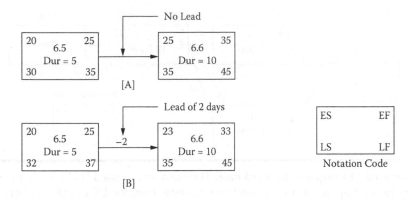

Figure 6.9   Network diagram with lead. A lead is a negative number on the arrow connecting two work packages that requires the successor work package to start earlier in the schedule, before the completion of the predecessor. The amount of advancement is equal to the lead number. In diagram [A], there is no lead. The results of the forward and backward passes are shown. In diagram [B], there is two days of lead, which is represented by a negative 2. Note in the forward pass, 25 (EF of 6.5) + (–2) = 23 (LS of 6.6). In the backward pass, 35 (LS of 6.6) – (–2) = 37 (LF of 6.5). Lead can be specified in any of the three precedent relationships: finish to start, start to start, and finish to finish. The rules for forward and backward passes always apply.

(numbers on the tops of the rectangles) was determined. Figure 6.11 demonstrates the backward pass for the same diagram as shown in Figure 6.10. Make sure you understand how the late schedule (numbers on the bottom of the rectangles) was determined. Figure 6.12 shows the forward pass, and Figure 6.13 shows the backward pass for a network diagram that contains every complication possible. Review the forward and backward pass procedures to make sure you understand how each day number is determined.

Analyze the finalized network diagram; pay particular attention to (1) total float, (2) free float, (3) critical path, and (4) estimated project duration. Chapter 9 explains three methods of increasing the project manager's confidence to complete the project on the network diagram's estimated finish day.

Validate the Network Diagram

Figure 6.14 shows the completed network diagram. Before proceeding to activity 3 through activity 5, we need to determine that the schedule represented by the network diagram is correct and do-able. Table 6.2 is a helpful checklist for this purpose.

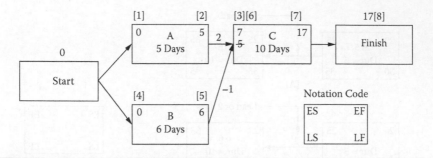

Figure 6.10 Example of forward pass. The numbers in brackets [ ] are not part of the network diagram. They are used here to show how the ES and EF numbers are determined. The number in [ ] from the network diagram above indicates the step in the procedure below that describes how the number is determined. [1] 0 (from Start) + 0 (on arrow) = 0 (ES of A). [2] 0 (from ES of A) + 5 days (duration of A) = 5 (EF of A). [3] 5 (EF of A) + 2 on the arrow = 7 (one alternative for ES of C). Because there are two arrows entering work package C, we must determine the second alternative before proceeding to determine the EF of C. We must determine the ES number that comes from the work package B path. [4] 0 (from Start) + 0 on the arrow = 0 (ES of B). [5] 0 (ES of B) + 6 day duration of B = 6 (EF of B). [6] 6 (EF of B) + −1 = 5 (second alternative for ES of C). There are two alternatives for the ES of C: 7 from step [3] and 5 from step [6]. On the forward pass, select the higher alternative; 7 is selected as the ES of C. [7] 7 (ES of C) + 10 day duration = 17 (EF of C). [8] 17 (EF of C) + 0 on arrow = 17 (duration time for the total project). This project is estimated to take 17 days to complete. The forward pass is complete when every work package has one number for each of the ES and EF corners.

Convert Schedule Day Numbers to Schedule Dates

Figure 6.14 shows the day numbers for the start and completion of each work package. The day numbers may be converted to calendar dates after the network diagram has been validated. Table 6.1 describes how to do this.

Normal work days[4]: This refers to the calendar of days for which the organization is open for business and the number of hours worked per day by the employees. For many organizations, normal work days include Monday through Friday and eight hours per day. Normal work days does not include those work days that fall on an organization-recognized holiday such as New Year's Day. It is important to determine the normal work days because you do not schedule work on non-work days such as holidays and weekends. Knowing the number of hours worked per day is also important because a 40-hour work package can be completed in five days if the employees work an eight-hour day. This same work package will require about five and a half days if employees work a seven-hour day.

Different categories of employees may have different normal work day schedules. In many organizations, employees who work in offices or laboratories work

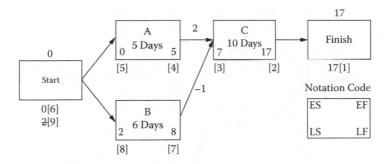

Figure 6.11 Example of backward pass. The numbers in brackets [ ] are not part of the network diagram. They are used here to show how the LS and LF numbers are determined. The number in [ ] from the network diagram above indicates the step in the procedure below that describes how the number is determined. [1] The number above the finish box is put under the finish box. [2] 17 (finish box) − 0 on arrow = 17 (LF of C). [3] 17 (LF of C) − 10 days duration = 7 (LS of C). [4] 7 (LS of C) − 2 days lag on arrow = 5 (LF of A). [5] 5 (LF of A) − 5 days duration = 0 (LS of A). [6] 5 (LS of A) − 0 on arrow = 0 under start. There are two alternatives for the number under start. Step 6 indicates 0 as one alternative. The other alternative is determined from step 9. On the backward pass, we keep the lower or lowest number. [7] 7 (LS of C) minus −1 on the arrow = 8 (LF of B). Remember from algebra: 7-(−1) = 8. [8] 8 (LF of B) − 6 days duration = 2 (LS of B). [9] 2 (LS of B) − 0 on the arrow = 2 (under the start). The backward pass is complete when every work package has one number for each of the LS and LF corners.

eight-hour days Monday through Friday. However, security personnel may work a different normal work day schedule; for these employees, it may be seven days a week including holidays, and each security person works four 12-hour days per week.

Analyze the Network Diagram

The definitions needed to analyze a network diagram include dependency relationships, total float, free float, critical path, lag, and lead.

Dependency relationships: The three dependency relationships FS, FF, and SS are described in Figures 6.2, 6.3 and 6.4, respectively. In the final network diagram, Figure 6.14, work package A and work package C are in a finish-to-start relationship. This means that the designer of the schedules wants C to start on the next working day after A is completed. (Remember that the Day Number Notation System in Figure 6.15 tells us that a work package starts on the day after the early start day number and ends on the early finish day number.) From Table 6.1 we can convert from schedule day numbers to schedule calendar dates. Work package A is scheduled to complete on day 5 (Friday, Nov. 9th) and C is scheduled to start

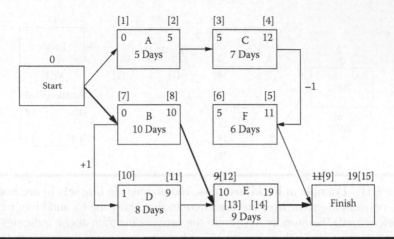

Figure 6.12 Forward pass with complex network. The numbers in [ ] are not part of the network diagram. They are used to show how the ES and EF numbers are determined. Before doing the forward pass, notice the relationships between pairs of adjacent work packages. The precedent relationships between start and A, A and C, B and E, F and finish, and E and finish are all finish-to-start (FS). For these relationships, the forward pass will always proceed from the EF of the predecessor to the ES of the successor; i.e., the procedures described in Figure 6.6 apply. The relationship between C and F is a finish-to-finish (FF) relationship. For this relationship, the forward pass will always proceed directly from the finish of the predecessor to the finish of the successor. Notice also that there are convergences at E and the finish box. Forward pass: [1] 0 (start) + 0 on arrow = 0 (ES of A). [2] 0 (ES of A) + 5 duration = 5 (EF of A). [3] 5 (EF of A) + 0 on arrow = 5 (ES of C). [4] 5 (ES of C) + 7 duration = 12 (EF of C). [5] 12 (EF of C) + −1 = 11 (EF of F). [6] 11 (EF of F) − 6 duration = 5 (ES of F). Note we have subtracted the duration to determine the early start of F. This is an exception to forward pass rule 1: always add. [7] 0 at start + 0 on arrow = 0 (ES of B). [8] 0 (ES of B) + 10 duration = 10 (EF of B). [9] 11 (EF of F) + 0 on arrow = 11 (finish). Step [15] will determine the alternative number coming from E. [10] 0 (ES of B) + 1 on arrow = 1 (ES of D). [11] 1 (ES of D) + 8 duration = 9 (EF of D). [12] 9 (EF of D) + 0 on arrow = 9 (ES of E). [13] 10 (EF of B) + 0 on arrow = 10 (ES of E). Choice is between 9 and 10. Rule 4 requires that we keep the larger number: 10. [14] 10 (ES of E) + 9 duration = 19 (EF of E). [15] 19 (EF of E) + 0 on arrow = 19 (finish). Choice is between 11 and 19. Rule 4 again requires that we keep the larger number 19.

on the morning of day 6 (Monday, Nov. 12th). Work package D is scheduled to be completed on day 9 (Thursday, Nov. 15th) and E is scheduled to start on the morning of day 11 (Nov. 19th). Work package E is scheduled to be completed on day 19 (Nov. 30th).

In the final network diagram, work package C and work package F are in a finish-to-finish relationship. If there were no number on the arrow connecting

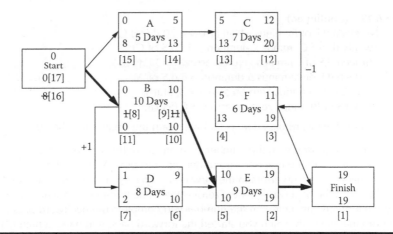

Figure 6.13 Backward pass with complex network. The numbers in [ ] are not part of the network diagram. They are used to show how the LS and LF numbers are determined. Before doing the backward pass, notice the relationships between pairs of adjacent work packages. The precedent relationships between start and A, start and B, A and C, B and E, D and E, F and finish and E and finish are all finish-to-start (FS). For these relationships, the backward pass will always proceed from the LS of the successor to the LF of the predecessor; i.e., the procedures described in figure 6.7 apply. The backward pass proceeds from the late start of the successor to the underside of start box. The relationship between C and F is a finish-to-finish (FF) relationship. For this relationship, the backward pass will always proceed from the LF of the successor directly to the LF of the predecessor C.

Notice also the convergence of two arrows upon B, one arrow touches the finish side of B and the second touches the start side. See steps [8] through [13].
Backward Pass:

1. Nineteen (on top of Finish) is transferred to under Finish.
2. Nineteen (Finish) minus zero on arrow = 19 (LF of E).
3. Nineteen (Finish) minus zero on arrow = 19 (LF of F).
4. Nineteen (LF of F) minus 6 duration = 13 (LS of F).
5. Nineteen (LF of E) minus 9 duration = 10 (LS of E).
6. Ten (LS of E) minus zero on arrow = 10 (LF of D).
7. Ten (LF of D) minus 8 duration = 2 (LS of D).
8. Two (LS of D) minus +1 on arrow = 1 (LS of B).

Notice there are two arrows converging upon B. We need to determine the two sets of LS and LF numbers for each arrow.

9. One (LS of B) plus 10 duration = 11 (LF of B).
     One and 11 are a set of numbers. There is another set.
10. Ten (LS of E) minus zero on arrow = 10 (LF of B).
11. Ten (LF of B) minus 10 duration = 0 (LS of B).

Zero and 10 are the 2nd set of numbers. Rule 4 of the backward pass directs to select the lower or lowest number or set of numbers. Therefore, 0 and 10 are retained.

Figure 6.13 (Continued)
12. Nineteen (LF of F) minus –1 on arrow = 20 (LF of C).
13. Twenty (LF of C) minus 7 duration = 13 (LS of C).
14. Thirteen (LS of C) minus zero on arrow = 13 (LF of A).
15. Thirteen (LF of A) minus 5 duration = 8 (LS of A).
16. Eight (LS of A) minus zero on arrow = 8 (Start).
17. Zero (LS of B) minus zero on arrow = 0 (Start).

Rule 4 of the backward pass directs that we retain the smaller number, 0.

The forward pass started with the number zero on top of the start box and the backward pass ended with the number zero under the start box. You may be confident that there are no errors in the forward and backward passes if the number you end with on the backward passes (zero) is the same number with which you started the forward pass (zero). If the number you complete the backward pass is different from the one with which you started the forward pass, you have definitely made one or more errors.

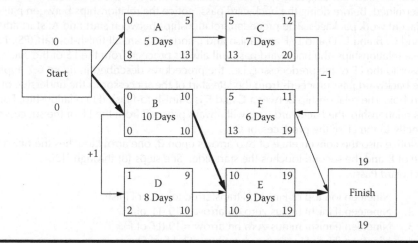

Figure 6.14 Final network diagram. This figure shows how the final network diagram would appear. Table 6.2 is a checklist intended to help assure that the network diagram is complete and do-able. Any problems identified by the checklist must be resolved before drawing the Gantt chart or other planning documents. The numbers in the corners of each work package indicate the day the package is scheduled to start and finish. For work package A, the early start (ES) is day 0; the early finish (EF) is day 5. The late start (LS) is day 8 and the late finish (LF) is day 13. (Refer to Figure 6.5 Day number notation system.) In accordance with the notation system, the early start of work package A is actually the morning of day 1 and the early finish is the evening of day 5. The late start is the morning of day 9 and the late finish is the evening of day 13. We will have to remember this when we convert the day numbers to calendar dates. Table 6.1 shows how to convert these day numbers into calendar dates.

Table 6.2 Network Diagram Checklist

1. Has the WBS been checked for completeness?
2. Is every work package listed in the WBS included in the network diagram with its own rectangle?
3. Does every work package have at least one arrow in and one arrow out, i.e., does each work package have at least one arrow point touching it and at least one arrow "tail" touching it?
4. Do all work package duration times include the effect of work package worker productivity and availability? (See activity 1 of this chapter.)
5. Have work package worker availabilities been negotiated and established?
6. Determine the critical path. Are you confident in the duration estimates for all the work packages on the critical path?
7. Do you have the people resources negotiated to staff each work package when it is scheduled to be performed?
8. The forward pass started with a zero above the start box and the backward pass ended with a zero under the start box.
9. Is the finish date acceptable to the client or sponsor?
10. We have the cash flow to fund the work packages in their current scheduling.
11. The ES, EF, LS, and LF day numbers or dates are indicated for every work package. There is only one number or date in each corner of each work package.
12. Work packages with long procurement lead times have been moved earlier in the schedule.

Note: Use the questions to validate the network diagram. Additional action is needed for any question receiving a "No" answer.

the finish of C to the finish of F, we would read it as requiring both work packages to complete on the same day. However, there is a –1 on the arrow between C and F. This means that the schedule designer wants F to complete one day earlier than C. Work package C is scheduled to complete on day 12 (Tuesday, Nov. 20th) and work package F is scheduled to be completed on day 11 (Monday, Nov. 19th).

Work package B and work package D are in a start-to-start relationship. If there were no number on the arrow, they would both start on the same day. However, the schedule designer wants D to start one day later than does B and therefore has put a +1 (a lag) on the arrow. Because of this, B is scheduled to start on day 1 (Monday, Nov. 5th) and D is scheduled to start on day 2 (Tuesday, Nov 6th).

Total float: Total float is the reserve of time a work package has that may be used (1) to delay the start of the work package from its early start day, (2) to expand the duration of the work package, or both (1) and (2), as long as the total amount

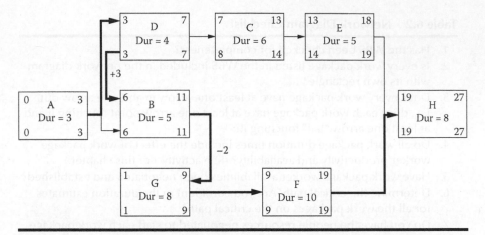

Figure 6.15 Explanation of critical path. (The start and finish boxes have been omitted because of space limitations.) Having zero total float is the first requirement for a work package being on the critical path. Work packages A, B, D, F, G, and H have zero total float. It is possible for a work package to have zero total float and yet not be on the critical path. For example, the sum of all the durations of the work packages with zero float including the three-day lag and the two-day lead is 39. All of the work packages cannot be on the critical path because the sum of the critical path durations must equal the project duration time. The project duration time is 27 days (the finish of H). Notice that the critical path touches two work packages but doesn't go through them: D and G. The critical path hits these work packages but bounces off like a ball bounces off a wall. The total duration of the critical path equals 27 if we exclude D and G from the critical path (3 duration of A + 3 lag + 5 duration of B + (−2) lead + 10 duration of F + 8 duration of H = 27). Notice that the critical path goes through A and B but the arrow between these two work packages is not highlighted. The arrow between A and B is not on the critical path because there are three days of free float on that arrow (6 ES of B − 3 EF of A); there is no float of any kind on the critical path. To verify this conclusion, add the duration along the path A, B, F, H; the total duration is 24 (3 duration of A + 5 duration of B + (−2) lead on arrow + 10 duration of F + 8 duration of H = 24). What does it mean that a work package is critical in the sense that it has no total float but it isn't on the critical path? Work package D has no total float (reserve time) and the free float at the end of the path (between finish of E and the start of H) is only one. If work package D comes in more than one day late, it will impact the start of H and thereby move the completion of the project by more than one day. That one day of free float is shared by work packages C and E also, so any delays of more than one day by these work packages will affect the project's final completion date. Work package G has no total float; if it completes later than day 9, the delay will impact the start of F, which is on the critical path. The implications of a work package having no total float but not being on the critical path can only be determined by studying the network diagram carefully.

of reserve time used does not exceed the total float available. The amount of total float a work package has is determined by:

$$\text{Total float} = \text{late finish} - \text{early finish} \qquad (6.10)$$

In Figure 6.14, work package C has total float of 8 (20 LF – 12 EF). The 8 days of reserve time may be used in three ways:

1. To delay the start of the work package from its early start day. Work package C is scheduled to start on day 5 but it becomes necessary to delay the early start by eight days. The duration is kept to seven days, so the late finish = 5 early start days + 8 total float reserve days used to delay the start + 7 days duration = 20, the latest acceptable finish day.
2. To expand the duration of the work package. Work package C starts on its scheduled early start day 5 but needs to use all of the total float reserve time to complete the work package. The start is day 5 + 7 days scheduled duration + 8 days of total float reserve used to complete the work package = 20, the latest acceptable finish day.
3. To delay both the start and expand the duration. Work package C is scheduled to start on day 5 but it is delayed by four days, so four days of reserve are used to delay the start. The duration of seven days is not adequate to complete the work package and the remaining four days of total float reserve are used. Start is day 5 + 4 days of reserve used to delay the start + 7 days scheduled duration + 4 days of reserve used to complete the work package = 20, the latest acceptable finish day.

Thus total float is a reserve of time similar to a military reserve on the battlefield. Just as a military reserve of troops may be committed to a single point or distributed to a number of points on the battlefield, the total float reserve of time may be committed to delay the start of a work package or expand the duration or both as long as you don't use more reserve of time than you have. A military commander with 200 troops in reserve may deploy them as necessary in whatever proportions necessary. However, the commander cannot deploy more reserve troops than the number that exist.

There is a further restriction on the use of the total float reserve. Total float is usually shared with one or more other work packages so any use of part or all of the total float reserve by a predecessor work package makes that used reserve unavailable to the subsequent work package. In Figure 6.14, work package A and work package C both have total float of 8 (13 LF – 5 EF) and (20 EF – 12 LF), respectively. The word "total" in "total float" is intended to remind us that the reserve is the total available to all work packages that share it. Work package A and work package C share the same eight days of total float reserve. If predecessor A uses some of the total float, that amount is automatically subtracted from the amount available to successor C. For example, if work package A needs to use four days of total float to expand the duration, the early finish date of A will change to 9 (0 ES + 5 duration + 4 days of total float = 9 EF). If the EF of A is 9, then the early start of C is also 9. The early finish of C is now 16 (9 ES + 7 duration = 16 EF). The total float of C is now only 4 (20 LF – 16 EF = 4).

Free float: Free float is the reserve of time that applies to a work package as the result of a convergence on the forward pass of two or more paths in the network diagram. Using up the free float does not impact the project's final delivery day, hence the name "free." Look for free float at convergences on the forward pass — where two or more arrow points come together; with two arrows, one path will have free float and the other will not. Arrows on the critical path (explained later) do not have free float.

Free float is determined by the following equations:

Free float = early start of the successor – early finish of the predecessor   (6.11)

or

Free float = project finish number – early finish of the predecessor   (6.12)

Equation 6.12 applies where there is a convergence at the finish box.

In Figure 6.14, there is a convergence on the forward pass at the start of work package E and another at the finish box. The arrow from the finish of B to the start of E has a free float of 0 (10 ES of E – 10 EF of B). The arrow from the finish of D to the start of E has a free float of 1 (10 ES of E – 9 EF of D). Notice that with two arrows, one has free float and the other does not. The arrow from the finish of F to the finish box has a free float of 8 (19 finish box – 11 EF of F). The arrow from the finish of E to the finish box has a free float of 0 (19 finish box – 19 EF of E). Again, one of the two arrows has free float and the other does not.

> **Tip:** Arrows on the critical path always have free float equal to zero because there is no float of any kind on the critical path.

How would I interpret free float? The crew performing work package D have a one day (free float) rest before starting work package E. The crew performing work package F can be released eight days before the project is completed if work package F completes on day 11 as scheduled, or they can expand the duration of F if needed by a full eight days and still not disrupt the project's scheduled completion day.

Notice also that free float occurs only on arrows that are in a finish-to-start relationship; it does not occur on arrows that are in a finish-to-finish or start-to-start relationship. For free float to be possible, there must be an arrow from the finish of the predecessor directly to the start of the successor; i.e., FS relationship. In the finish-to-finish and start-to-start relationships, this arrow doesn't exist.

Free float is the result of unused total float. The amount of total float used automatically reduces the free float by a similar amount. In Figure 6.14, work package D has total float of 1 (10 LF of D − 9 EF of D). The free float on the arrow from D to E is also 1 (10 ES of E − 9 EF of D). If the crew performing work package D needs to use the one day of reserve time (total float) to complete the work package, then the EF day is 10 not 9 (1 ES + 8 duration + 1 total float reserve). If the EF of D is 10 and the ES of E is 10, then the free float on the arrow is 0 (10 ES of E − 10 EF of D).

> **Tip:** Free float is not indicated on the network diagram because a positive number on the arrow would be read as a lag and the notation FF 1 on an arrow would be read as a finish-to-finish relationship with a one-day lag.

Critical path: In a network diagram:

1. The critical path is the path that connects all the work packages (with one exception) that have zero total float. (See Figure 6.15.)
2. The total duration of the critical path is the shortest time in which the project may be completed.
3. The total duration of the critical path is the longest duration of all the paths in the network.
4. The duration of the critical path is the duration of the project.
5. There is no float (total or free) on the critical path.

The critical path is emphasized by highlighting the arrows connecting the boxes or the boxes or by color, etc. Figure 6.15 shows a complex network diagram that requires careful reading to determine the critical path.

Lag and lead are described in Figure 6.8 and 6.9. When reading a network diagram, be careful to note where the lags and leads appear. These adjustments to the schedule to delay the start of a successor (lag) or to start a successor before the predecessor completes (lead) are requirements established by the schedule developer. The duration of lags and leads are part of the cumulative project duration if they appear on the critical path as they do in Figure 6.15.

Activity 3: Draw the Gantt Chart and Milestone Chart

Gantt Chart

The Gantt chart is a picture of the project over time; it shows the planned start and end day number (or date) for every work package. It may also show the actual start and end day numbers. All Gantt charts have a similar format. Time is always shown on the horizontal axis with work package designations on the vertical axis. Commercial software provides numerous variations of the way work packages are represented. The network diagram must be completed before the Gantt chart can be drawn because it summarizes the information on the network diagram.

One common format is to show the work packages as rectangles with the left edge of the rectangle representing the start of the work package and the right edge representing the finish. Sometimes the total float associated with each work package is shown as a "tail" extending from the finish edge to the late finish day. Figure 6.16 describes how to draw the Gantt chart; Figure 6.17 is an example showing the planned schedule and the actual schedule.

Resource Gantt Chart

Sometimes it is necessary to have a resource Gantt chart. This lists the occupational specialty codes (job titles) needed to perform the work packages. This information is useful to functional managers who must allocate the people resources to accomplish the schedule. Figure 6.18 is a partial example. Figure 7.1 shows another version of the resource Gantt chart.

Milestone Chart

A milestone is an important event with a date that the project manager, sponsor, or customer want to monitor. There are three kinds of milestones:

1. Any significant event such as the submission of a deliverable.
2. A level of effort milestone, which indicates the date on which the project has consumed a planned number of hours of labor. For example, as of July 1, 2008, we expect to have consumed 10,000 cumulative hours of labor on the project.
3. A funding milestone is similar to the level of effort milestone; it indicates the date on which the project expects to have spent a cumulative amount of money. For example, as of July 1, 2008, we expect to have spent, cumulatively, $500,000.

A level of effort milestone is of interest to human resource management because it is necessary to have the capacity (numbers of employees) to expend the 10,000 of labor. A funding milestone is of interest to financial managers because it is necessary to have the cash flow to sustain this level of expenditures. The most common milestone is one that addresses a significant project achievement. The completion of the project is an obvious milestone. Two common ways to document milestones

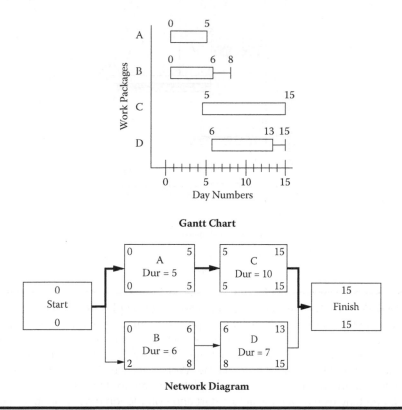

Figure 6.16   Drawing the Gantt Chart. Instructions for drawing the Gantt chart: (1) Complete and check the network diagram. See Table 6.2. (2) Draw the horizontal time scale for the Gantt chart. The scale must accommodate the project duration. In the example, the scale must be from 0 to 15 — the project duration. The time scale may be in day numbers or calendar dates. (3) List all the work packages on the vertical axis. List them in any order you require. They may be listed with A at the bottom and proceed upward; they may be listed with A at the top and proceed downward. You may list them in their approximate chronological order starting with the earlier work packages. You may list them in groups where each group represents a function or department or control account. With this last method, you are drawing the Gantt chart in a format similar to the work breakdown structure. (4) Draw a rectangle for each work package with the left edge of the rectangle at the early start number shown in the network diagram. Draw the right edge of the rectangle at the early finish number. For example, work package A has an early start day of 0 and an early finish day of 5. Do this for all work packages. Always use the early start and early finish numbers to draw the rectangles for each work package. (5) If you wish, you may add a "tail" to each work package rectangle to represent the total float. Work packages B and D have total float of 2 each. Draw the tail so it extends to the late finish number. Work package B has a tail that extends to day 8 and D has a tail that extends to day 15. Work packages without a tail have no float; they are critical. (6) If you wish, you may show the work packages with the same total float adjacent to each other with a dependency arrow to indicate that one work package is the predecessor to the other. The arrow point indicates the successor. See Figure 6.17 for an example. (7) You may include the actual start and completion dates on the same Gantt chart as the planned start and completion dates or you may put the actual on a different chart. See Figure 6.17 for an example.

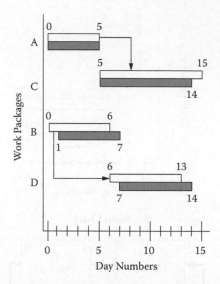

Figure 6.17   Gantt Chart showing plan versus actual plus dependencies.
This Gantt chart shows work packages represented in dependency order with a
dependency arrow to indicate which work package is the successor. The arrow
point touches the successor. Work package C is the successor to work package A;
for C to start on time, A must end on time. Work package D is dependent upon
B because B, the predecessor, must complete on time for D to start on time. The
shaded rectangles indicate the actual start and completion days: work package A
started and ended on time. Work package C started on time, day 5, and ended
early — on day 14. Work package B started one day late and ended one day late.
Work package D started on day 7, one day late, and finished on day 14, also one
day late.

is to include them on the Gantt chart or to list them on a work processor document
of milestones. See Figure 6.19 and Table 6.3.

Activity 4: Cross-Check the Schedule against
the Resource Plan

Assure that the resource plan (and the resource Gantt chart) has the allocated
resources needed to accomplish the schedule. Cross-check the network diagram
against the resource plan to assure that people and other resources will be available
to support the schedule as depicted in the network diagram (or the Gantt chart).
Get a commitment from functional managers to provide their people in accordance

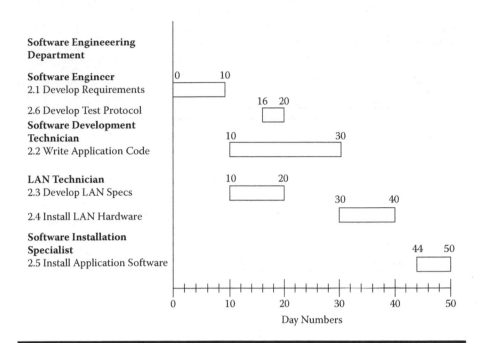

Figure 6.18 Partial resource Gantt. This chart shows the occupational specialties performing the work packages.

with requirements listed in the resource plan. Revise the resource plan and the network diagram (and Gantt chart) as necessary to match resource availability if necessary. If the project is to be completed by the project team, eliminate or minimize those times in the schedule where there is too much work or not enough work for the number of people on the team.

Activity 5: Get the Schedule (Network Diagram and Gantt Chart) Approved

When briefing the schedule to upper management and sponsors, be sure to stress the resource requirements associated with the schedule. The resource Gantt chart would help to accomplish this. Be sure the managers who own the resources know the levels of support needed to keep the project on schedule. The purpose of cross-checking the resource plan with the schedule (activity 4) was to negotiate the people resources needed to achieve the schedule, so the briefing should be a somewhat anticlimactic sharing of information because negotiations with the functional manager and decision making have already been completed.

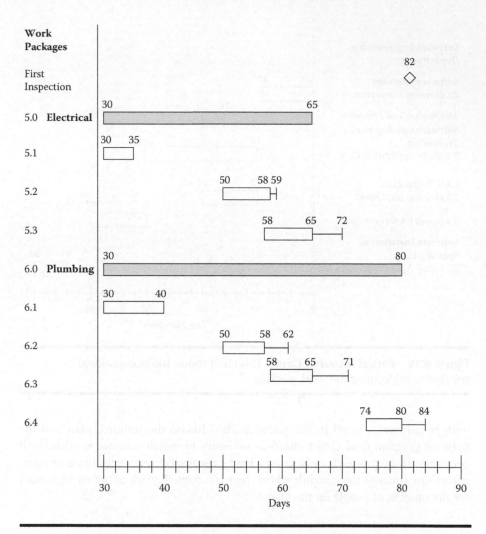

Figure 6.19 Partial Gantt Chart showing a milestone. This is a partial Gantt chart indicating the electrical and plumbing work packages. It shows only the planned start and end dates, not the actual start and end dates. The first building inspection on day 82 is shown as a planned milestone. An actual milestone would be shown as a shaded diamond. (Work packages without tails are on the critical path.)

Table 6.3   Example of Milestone Chart in Word Processor Format

MILESTONES FOR CONSTRUCTION OF THE HOUSE AT 2182 MADISON DRIVE

Milestone	Planned Date	Actual Date
Begin site development	March 1, 2008	
First bldg inspection	April 15, 2008	
Final bldg inspection	June 24, 2008	
Occupancy permit	June 25, 2008	
Customer walk-through	June 28, 2008	
Real estate closing	July 1, 2008	

Note: The planned milestone dates are shown. The actual dates remain blank until the milestone is achieved.

Notes

1. "Adding Risk into Project Estimates — PERT vs. Monte Carlo," by Guy L. De Furia, Ph.D., PMP. (Article first appeared in the April 2007 issue of *ESI Horizons.* © ESI International, 901 North Glebe Road, Suite 200, Arlington, VA 22203; www.esi-intl.com. With permission.)
2. Veracity demands the clarification: Booz-Allen was the consulting firm assisting the Navy at the time.
3. Increasing your confidence in the estimated effort times for work packages is covered in "Measure Schedule Risk Using Standard Deviation," by Guy L. De Furia, Ph.D., PMP. (Article first appeared in the January 2002 issue of *ESI Horizons.* © ESI International, 901 North Glebe Road, Suite 200, Arlington, VA 22203; www. esi-intl.com. With permission.)
4. Working time and work calendar are other terms that translate to normal work days.

Chapter 7

Project Resource Plan

Mega Recipe for Developing the Resource Plan

Activity 1: Develop the resource plan.
Activity 2: Validate the resource plan.
Activity 3: Finalize the resource plan.

When I was a teenager my father was a general contractor who built houses. During summer vacations from school I worked with him. One Monday, I got in the pickup truck with him and drove to the site where three houses were being constructed. The basement walls had been completed the previous week, and my father had hired three carpentry crews. When we arrived at the site, there were the carpentry crews drinking coffee and eating donuts, as is typical for construction people. Immediately my father noticed that there was no lumber with which to construct the three houses. He got into the truck and drove to a telephone (cell phones hadn't been invented yet) and asked about the lumber. He was told by the yard foreman that the lumber was being loaded on the truck as they spoke, and that the best they could do would be to promise that it would be delivered by the end of the day. Pop had no choice but to tell the carpenters to go home and, yes, they would be paid for this day, and to please come back tomorrow. It wasn't until many years later when I was running a project in the execution phase that I realized that much of my work involved "making sure the right person shows up at the right place, at the right time, with the right understanding of what needs to be accomplished, with the right tools, equipment, and material with which to be successful." Any break in this chain of necessary conditions will cost the project time and money, and frequently both. Pop had everything arranged except for the lumber. This one coordination failure cost him 12 days of carpenters' salaries and 12 person-days of work lost.

To be able to do the assertive advanced coordination necessary to prevent this from happening, the project manager needs a spreadsheet that shows all the resources needed for each work package. Notice that the right person not showing up at the right place, etc., can occur on any work package. Assertive advanced coordination means getting on the telephone two weeks before a work package is supposed to start and making sure that "the right person knows where to be and has a complete understanding of what has to be accomplished, has the tools, material, etc., with which to be successful."[1] Advance coordination should be done at least two weeks before each work package is to start because the project manager will need time to find a substitute worker if the original worker is not available (promoted, transferred, on vacation, etc.) or to coordinate the arrival of the material, etc.

The document that contains the information the project manager needs to do this advanced coordination is the project resource plan. Table 7.1 shows an example. Notice that every work package must be shown in the first column. The other columns are used to indicate the resources needed to perform the work packages. Be sure to indicate whether the work package will be performed on a full-time basis or less-than-full-time basis. If other than full-time, indicate percent of availability.

Table 7.1 Project Resource Plan Template

			PROJECT RESOURCE PLAN			
Work Package	Start & End Dates	Who	Budget	Special Equipment	Material	Comments (e.g., full- versus part-time[a])

[a] Comments can indicate full- versus part-time effort as a percent number, e.g., three-quarters time is expressed as .75 or 75%. For instructions, see activity 1.

Example 1: A worker estimates it will take her 40 hours to accomplish the work package but she will have to contribute only 50% of her time to this work package. This tells the project manager that the project will wait for 80 hours to get the completed work package although the worker will charge the project for 40 hours. (See examples 1 to 4 in Chapter 6.)

Activity 1: Develop the Resource Plan

Before you can develop the project resource plan, you must have a complete work breakdown structure (WBS) down to the work package level and the project schedule (precedent diagram). The procedure is as follows (see Table 7.1.):

Step 1: Complete the WBS. Chapter 4 describes how to develop the WBS. List all the work packages from the WBS on the resource plan in the "Work Package" column.

Step 2: Complete the precedent diagram for the project; it is this document that determines the early start and early end date for each work package. For each work package enter these dates in the "Start/End Dates" column.

Step 3: Determine the person, department (e.g., engineering), or subcontractor company that is going to perform the work package. Determining who will perform each work package involves answering four questions:
   a. What expertise (experience or training) is needed to perform the work package?
   b. Who has the capability to perform the work package?
   c. Of the capable people or companies, who is available to perform the work (when it is scheduled to be accomplished)?
   d. Which of the capable and available people or companies would do the work package at the best cost?
   These four questions must be answered for every work package. Indicate the person, department, or company in the "Who" column.

Step 4: Enter the budget for each work package. The procedure for determining the budget for each work package is discussed in Chapter 5.

Step 5: If a work package requires special equipment, i.e., equipment not normally required, enter it in the "Special Equipment" column. For example, a work package is titled "Install Air Conditioner." Plumbers who would do this installation have the tools they need. However, this air conditioner must be installed on the roof. This installation will require a crane. Putting "crane" on the resource plan reminds the project manager to coordinate the crane's availability on the right day.

Step 6: Enter the materials or supplies needed to perform a work package in the "Material" column. Use an abbreviation or reference to another document to indicate the list of materials. For example, listing the lumber needed to construct a house on the resource plan is not feasible. The abbreviation "BOM" (bill of materials) reminds the project manager that a bill of materials exists and that this material is needed to perform the work package.

Step 7: Enter any comments relevant to each work package in the "Comments" column. These serve as reminders to the project manager. For example, the work package titled "Install and Spackle Wall Board" is scheduled to be performed during the last two weeks of January. A comment for this work package, such as "Room temperature must not be below 45 degrees F" reminds the project manager to make sure space heaters are available during the last two weeks of January.

The resource Gantt is useful in displaying the people from each department of the organization that are scheduled to support the project. This type of Gantt chart makes it easy for functional managers (e.g., chief of engineering in the example) to identify their total commitment to the project. Figure 7.1 is an example of a partial resource Gantt.

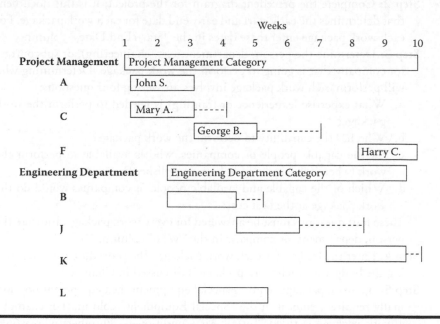

Figure 7.1 Partial resource Gantt. Work packages are grouped by function or department. The Gantt chart shows the individual project team member responsible for each work package. The individual who will perform each of the engineering work packages is left blank because the decision of who is assigned each work package is made, in this example, by the chief of engineering. The tail behind the work package indicates the amount of total float. Work packages without float are on the critical path. Figure 6.18 shows another version of a resource Gantt chart.

Activity 2: Validate the Resource Plan

Sometimes a project has a full-time staff. When this is the case, there is the added concern that there may be times during the project execution and closeout when the project does not have enough work scheduled to keep the team members productively employed, and there may be times when there are not enough people to accomplish all the work scheduled. Releasing team members to their functional departments is not a good idea because it may be difficult to get them back or the functional manager may substitute someone else.

Workload leveling has as its purpose the elimination of those times during the project when there is too much or not enough work scheduled for the number of people on the team. For example, the full-time project staff consists of five people. The project team is experienced in the technology of the project, and team members are essentially interchangeable. One person may substitute for another. Workload leveling identifies those times for which we have scheduled too much work or too little work for five people to accomplish. The resource loading matrix in Table 7.2 shows an example of this problem. The requirements for people during days 6 through 15 exceed the five people that are on the team. The matrix in Table 7.3 shows the fixed schedule after the procedure called workload leveling has been applied.

To identify that your project has a resource loading problem and to bring about a solution require a "before" loading matrix, which will show where the problem exists, and an "after" matrix, which will show how the problem has been resolved. The procedure requires a completed precedent diagram for the project similar to the example in Figure 7.2.

To construct the "before" resource loading matrix, follow the five steps below. (Grid paper or spreadsheet software would be useful here.) Refer to Table 7.2.

Step 1: List all work packages on the vertical axis. Get this information from the WBS.

Step 2: Indicate the time scale on the horizontal scale. The time scale starts with the number 1 and ends on 25, the last day of the project. The last day is shown in the precedent diagram; it is the number over the finish box.

Step 3: In the body of the matrix, indicate the numbers of people working on each work package during the time it is scheduled to be performed. The precedent diagram in Figure 7.2 shows that work package A starts on day 1 and ends on day 5. Remember the convention: when the precedent diagram starts with 0, as it does in Figure 7.2, it means that the early start number in the work package is the day before the work package starts, and the number in the early finish corner means the day the work package ends. In Figure 7.2, work package A has a 0 in the early start corner, which means it actually starts on the morning of day 1. It has a 5 in the early finish corner, which means that it is planned to finish at the end of

Table 7.2 Resource Loading Matrix BEFORE Leveling

	1	2	3	4	5	6	7	8	9	10	11	12	13	14	15	16	17	18	19	20	21	22	23	24	25
H	-	-	-	-	-	-	-	-	-	-	3	3	3	3	3	-	-	-	-	-	-	-	-	-	-
G	-	-	-	-	-	-	-	-	-	-	-	-	-	-	-	-	-	-	-	-	-	-	-	-	-
F	-	-	-	-	-	-	-	-	-	-	-	-	-	-	-	-	-	-	-	-	-	-	-	-	-
E	-	-	-	-	-	2	2	2	2	2	2	2	2	2	2	-	-	-	-	-	-	-	-	-	-
D	-	-	-	-	-	2	2	2	2	2	-	-	-	-	-	-	-	-	-	-	-	-	-	-	-
C	-	-	-	-	-	2	2	2	2	2	-	-	-	-	-	-	-	-	-	-	-	-	-	-	-
B	1	1	1	1	1	-	-	-	-	-	-	-	-	-	-	-	-	-	-	-	-	-	-	-	-
A	2	2	2	2	2	2	2	2	2	2	2	2	2	2	2	4	4	4	4	4	2	2	2	2	2
Day	1	2	3	4	5	6	7	8	9	10	11	12	13	14	15	16	17	18	19	20	21	22	23	24	25
Loading	3	3	3	3	3	8	8	8	8	8	7	7	7	7	7	4	4	4	4	4	2	2	2	2	2

Note: Team has five full-time members. Work scheduled exceeds the team's capacity on days 6 through 15. There is too little work scheduled on days 1 through 5 and 16 through 25. For instructions, see activity 2.

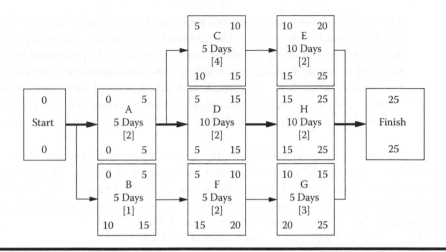

Figure 7.2   Example of precedent diagram. The number of team members needed to perform a work package is shown in [ ]. For example, F needs two team members working for five days to complete the work package.

day 5. Likewise, work package C is scheduled to start on day 6 and end on day 10.

In Figure 7.2, work package A had two people working on it for five days, so enter the number 2 in the matrix for days 1, 2, 3, 4, and 5. The number 2 is shown in days 1 through 5 because the precedent diagram shows it starting on day 1 and ending on day 5. Table 7.2 shows this. The precedent diagram (Figure 7.2) shows that work package B needs one person working on it during days 1 through 5. On the matrix diagram, for work package B, enter the number 1 for days 1 through 5. The precedent diagram shows that B has float through day 15, so draw the tail (" -]") to represent the float. Each dash (-) represents a day. The dashes line up with days 6 through 15.

Work package D requires two people; it starts on day 6 and ends on day 15, and there is no float because the early finish and the late finish are both day 15. The matrix therefore has no tail for D. In the loading matrix, enter a number 2 for days 6 through 15. Continue this process of listing the numbers of people needed to work on the work packages; be careful to show the float where it exists.

Step 4: Add the numbers in each column and enter the number in the "Loading" row. In Table 7.2, days 1 through 5 need three people each day. Days 6 through 10 require eight people. Loading refers to the number of people needed to perform the work scheduled for each day.

Step 5: Determine if and when the people loading problem exists. In Figure 7.2, we see that the original schedule calls for three people for days 1 through 5 — not enough work for the entire team unless the project leader has other work (not related to a work package) that will need to be done. The original schedule reflected in Table 7.2 shows a requirement for eight people in each of days 6 through 10; this is more work than can be performed by the team of five people. Reading the "Loading" row identifies those times during the schedule when there is too much work scheduled and those times when there is not enough. The resolution of this resource loading dilemma is called workload leveling. (Resource leveling is another name for workload leveling.)

Workload leveling seeks to eliminate those times when there is too much work or too little work scheduled for the number of people on the team. All the information needed to do this is shown on the loading matrix. It requires that you use the information in the "before" matrix (Table 7.2) to create the solutions that are shown in the "after" matrix similar to the one in Table 7.3.

There are three basic strategies to workload leveling:

1. **Shifting work within its float:** Chapter 6 introduced the idea that there are work packages that have total float within which the work package must be performed. Total float equals the late finish minus the early finish. In Figure 7.2, work package B has 10 days of total float (15 − 5 = 10). This means that although B is originally scheduled to be performed on days 1 through 5, it can be rescheduled anywhere between day 1 and day 15. That is, we may shift the entire work package to a later start and end date as long as we do not reschedule it to end later than day 15. When we do workload leveling, this is one strategy we may use to reduce the total amount of work in some days and increase it in later days. Shifting work always moves work later in the schedule; you cannot shift work forward or earlier in the schedule. The "after" matrix shown in Table 7.3 shows that work package E was shifted within its float; it was originally scheduled to be started on day 11 and rescheduled to start on day 16.

2. **Interrupting work within its float:** The nature of some work allows it to be interrupted without causing any problems. For example, the pouring of concrete may not be interrupted once it is commenced because it sets (gets hard) quickly. Alternately, the writing of a report or the collection of data may be interrupted with no harm done to the effort. The attribute of a work package that allows it to be started and then suspended for a time and then restarted allows us to use the interrupting strategy. In Table 7.3, F and G were interrupted. This is easy to see because interrupted work packages always have dashes ("-") in the middle. Work package F and work package G were also stretched, which is the third strategy.

3. **Stretching work within its float:** Some work packages involve more than one person to perform. For example, the collection of data may be planned as a four-person effort that requires one week to do and there is one week of float. Think of this work package as a thick elastic band, four people thick

Table 7.3 Resource Loading Matrix AFTER Leveling

	1	2	3	4	5	6	7	8	9	10	11	12	13	14	15	16	17	18	19	20	21	22	23	24	25
H																2	2	2	2	2	2	2	2	2	2
G	1	1	1	1	1											2	-	-	-	1	2	2	2	2	2
F						1	1	1	1	1	[-	1	1	1	1	-	-	-	2	2	2	2	2	2	2
E																2	2	2	2						
D						2	2	2	2	2	2	2	2	2	2										
C						2	2	2	2	2	2	2	2	2	2										
B	1	1	1	1											-]										
A	2	2	2	2	2																				
Day	1	2	3	4	5	6	7	8	9	10	11	12	13	14	15	16	17	18	19	20	21	22	23	24	25
Loading	3	3	3	3	3	5	5	5	5	5	5	5	5	5	5	5	5	5	5	5	6	6	6	6	6

Note: Work scheduled equals the capacity of the five-person team with the exception of days 1 through 5. Work package C was stretched; E was shifted within its float; F and G were interrupted and stretched. Work packages A, D, and H could not be changed because they have no float. Work package B could have been moved but to do this would not have served any useful purpose. The only remaining problem is in days 21 through 25 when the workload requires six people.

and five days long. If we stretch an elastic band it gets longer and thinner. Stretching this work package might require two people to perform the work package and take 10 days to do it. Stretching a work package is do-able as long as the duration falls inside the late finish date. Remember that total work effort equals the number of people times the number of days. The original schedule for work package C in Table 7.2 requires 20 people days of effort (4 people × 5 days). The stretched schedule requires two people to work 10 days; this equals the original 20 days of effort, only now it is stretched over 10 days instead of five. In the "after" matrix, Table 7.3, work package C was stretched; F and G were both stretched and interrupted.

The result of the workload leveling is shown in Table 7.3. The workload for days 6 through 20 has been adjusted to 5. The workload for days 1 through 5 remains at 3. The workload for days 21 through 25 exceeds capacity by 1 per day.

The following "desperation" strategies may be tried if the basic strategies 1 through 3 listed previously don't solve all the people-versus-workload problems:

1. Move the final completion date to a later date; this will provide more float within which to use the basic strategies.
2. Seek temporary help from within the organization or from temporary employee companies, or work the team members so they put in longer days to make up for those times when there are not enough people to do the work package. This will not make the project manager popular!
3. Redraw the precedent diagram; that is, change the chronological order of the work so you avoid some of the people-versus-workload problems. Oftentimes, the results of workload leveling will significantly improve the capacity-versus-workload situation but not completely solve it.

Generic Responsibilities Many projects have generic responsibilities that must be delegated and documented as team member monitorship or additional duty assignments. Many are common to all projects; some exist under unusual circumstances. Most project managers must assign particular team members, for example, to monitor the schedule (network diagram and Gantt charts), perform liaison with the procurement department, monitor the expenditure of money against the budget, administer the scope change control procedure, perform the project cost and schedule earned value calculations, monitor and administer the risk management plan, or establish and maintain the project filing system. On some occasions the project manager may need to assign responsibility for less frequently encountered areas such as physical or personnel security, site management, housing and feeding of project personnel, medical support, transportation, or hiring of local workers. All areas of responsibility must be delegated, and the delegation must be clear and documented. Sharing who-is-responsible-for-what with all team members facilitates teamwork.

Activity 3: Finalize the Resource Plan

The resource plan is an important document with which to manage the project. However, its utility is directly connected to the degree to which it represents firm commitments. Finalizing the resource plan means getting firm commitments from functional managers to provide the numbers of skilled people at the time called for in the plan.

When negotiating the resource plan with a functional manager, be sure to share the resource Gantt chart and the work package work orders. These documents detail the commitment being negotiated.

Note

1. *Source Project Management (Advanced) for DAC,* published by ESI International, 901 North Glebe Road, Suite 200, Arlington, VA 22203, October 2006, pp. 5–7. With permission.

Chapter 8

Project Filing System

Mega Recipes for Managing Project Information

Information storage standards and methods vary widely among organizations. This chapter suggests an approach to setting up a project filing system.

    Activity 1: Develop the system of files.
    Activity 2: Retain important information.
    Activity 3: Dispose of and archive information.

Purpose of the Filing Activities

The purposes of these activities are to develop a filing system that:

1. Facilitates the storage and easy access to the information gathered, generated, and used in the course of the project
2. Provides the information base with which future estimates of time, cost, resources, and risks may be more accurately determined
3. Documents actual cost and time performance
4. Encourages the gathering and sharing of lessons learned with which organizations improve their ability to conduct projects

Role of the Project Manager

Many organizations have standard operating procedures (SOPs) for the storage and archiving of information. But formalized project management is relatively new to many organizations that have not yet developed these procedures. In the absence of these organizational procedures:

1. The project manager is responsible for creating a system to store, retrieve, share, and dispose of information during the entire project life cycle. The project manager also anticipates the organization's future needs for information about the current project, and sees that this information is archived. Estimating time or cost to perform work packages is an example. Organizations will not improve their ability to estimate time and cost unless they organize and retain the estimates and actual time and cost for the current project.
2. The project manager is also charged with assuring that the actual performance of each work package and the total project are recorded and retained because this information is crucial to future estimating. Recording actual performance may be done by anyone on the team including work package managers or project managers as an adjunct to monitoring the project.

Problems Managing Information

1. Not understanding the purposes of a filing system and the benefits, there are two reasons why this topic may be given little attention. Organizations that do not keep records of estimates and records of actual performance do not improve in their ability to estimate the cost and time to perform work.
2. Information management depends heavily upon the actions of recording what is planned and what actually happened. The planning process automatically includes the creation (recording) of the planning documents but recording what actually happened requires a special and separate effort from the execution of the plans. Recording actual performance will be haphazard unless clear expectations for doing so are set and reinforced by the project manager.

Activity 1: Develop the System of Files

Every project must have a system that provides for easy storage and retrieval of information. The system must be simple and clearly connected to the major project activities. Following are the file names and their contents (see Table 8.1 for a list of file topics):

1. **Project definition:** The need and problem assessment necessitating the project should be stored in the project definition file. The business case definition for the project as well as notes, comments, clarifications,

Table 8.1 List of Project Files

(1) Project definition	(8) Risk management plan	(15) Lessons learned
(2) Contract	(9) Scope changes	(16) Final risk assessment
(3) Charter	(10) Closeout plan	(17) Final project evaluation
(4) WBS and WP work orders	(11) Purchase orders and subcontractor closeout	(18) Contract and client closeout
(5) Estimates & control limits	(12) Project closeout	(19) Personnel and team closeout
(6) Budget & cost baseline	(13) Project performance	(20) Organizational closeout
(7) Schedules	(14) Baseline changes	(21) Project final report

and decisions that occurred during the business case briefing should be recorded and retained in the project definition file also.

2. **Contract and proposal:** The documentation of a customer solicitation and successful proposal and contract are retained in the contract file.
3. **Charter:** The project charter should be retained in its own file. The charter is the document that approves the project for funding, identifies the project manager, and outlines the responsibilities and authority of the project manager.
4. **WBS and work package work orders:** The work breakdown structure (WBS) lists all the work (work packages) that must be performed during the project. The work package work orders include all the detailed information about each work package including the cost and duration baselines and the control limits for each.
5. **Estimates and control limits:** This file includes the work package estimating work sheet (Figures 5.2 through 5.4) and the work order (Figure 4.1 and Figure 13.1) for each work package. The work order describes the scope of each work package plus its baseline, and upper and lower control limits of cost and duration time. It includes the documentation of the calculations with which control limits were determined. (See activities 2 and 3 in Chapter 10 and Table 10.1, Table 10.2, Figure 10.1, and Figure 10.2.)
6. **Budget and cost baseline:** This file includes the documentation on the bottom-up estimate of project cost. (Refer to Table 5.5 and Table 5.6.)
7. **Schedule:** This file contains the network diagram, the Gantt chart, and the milestone chart.
8. **Risk management:** This file includes the risk management plan plus documentation on the efforts to manage and control risks plus the consequences of these actions.

9. **Closeout plan:** This file contains the details of the activities and subtasks required to complete the project closeout.
10. **Scope changes:** This file includes the scope change procedure plus the documentation of suggested changes, estimated impact on cost and time, and their final disposition.
11. **Procurement and subcontractor closeout:** Included in this file are all communications, purchase orders, and contracts with subcontractors needed by the project. Create a file for each subcontractor if necessary.
12. **Project performance, evaluation, and forecasting:** Store the earned value status reports (CV, SV, CPI, SPI) for the work packages in this file. Also include the actual duration time and cost versus the baseline, upper and lower control limits of time and cost. (Refer to Table 10.3, Figure 10.3, Figure 10.4, and Figure 13.2.)
13. **Baseline changes:** Evaluating project performance against its objectives and terminal scope, time, and cost expectations may make it necessary to re-baseline the project. The documentation about the need to change the baseline and the readjusted expectations is retained in this file.
14. **Lessons learned:** Lessons learned are bits of experience that help project teams maintain good performance or improve performance. These lessons are surfaced when we ask three questions:
    a. What went well?
    b. What didn't go well?
    c. What should we do differently next time?
    This file retains this information. (See Figure 12.6.)
15. **Final risk assessment:** There are opportunities and threats unique to the closeout of a project. This file retains these opportunities and threats so the team or the organization may act upon these as necessary. Their recognition and the actions taken also serve to generate lessons learned. (See Figure 12.5.)
16. **Final project evaluation:** This file retains the information about the final judgment of the success of the project. Did the project accomplish its goals and objectives? Did it complete on time or early or late and was the duration time within acceptable variance? Did the project complete on budget and was the variance within acceptable variance? This file also documents the efforts to determine if all the work on the project was completed. If there are exceptions, the reasons for the exceptions are noted. Memos that accompany deliverable submissions and acceptance memos from the customers are also retained in this file. (See Figure 12.1, activity 1 in Chapter 14, and Figure 14.1.)
17. **Contract closeout:** This file documents that all the requirements of the contract have been met.
18. **Personnel closeout:** This file contains all the personnel records executed during the project as well as memos of appreciation for those who worked on the project. (See activity 7 in Chapter 14 and Figure 14.3.)

19. **Organization closeout:** Documentation related to return of unused funds, release of equipment borrowed from other parts of the organization, and notification of room availability to the space management office are the kinds of administrative tasks that are retained in this file. (See activity 3 in Chapter 14 and Figure 14.4.)

Activity 2: Retain Important Information

With a filing system in place, the project manager needs to set clear expectations about its use and reinforce behavior consistent with the four purposes of the filing system.

Activity 3: Dispose of and Archive Information at the Conclusion of the Project

The format and length of time that information is retained varies from organization to organization. Future utility is the criterion for deciding:

1. What information should be kept?
2. How long should information be kept?
3. In what format (paper, electronic, or Web-based) should the information be kept?

Legal considerations also play a role as well as the possibility of an Internal Revenue audit. The legal considerations to archiving can best be determined by the organization's legal staff.

However, the very purposes of the filing system would indicate which files should be retained indefinitely. The purposes of the system are to provide information with which to improve the organization's ability to estimate time and cost of work packages, identify risks and potentially useful risk-reducing strategies, provide information on actual time and cost to perform work packages, and to retain lessons learned with which to improve the organization's ability to conduct projects. The following files should be retained indefinitely in a format easily accessed by future project teams:

1. WBS and the work orders
2. Estimates and control limits
3. Budget and cost baseline
4. Risk management
5. Lessons learned

Chapter 9

Risk Management Plan

Mega Recipes for the Risk Management Plan

The risk management plan is the team's roadmap for dealing with the threats to and opportunities for the project. Developing this plan includes the eight activities listed below. A conscientious performance of these activities will produce a professional risk management plan.

Activity 1: Prepare the team to do risk management.
Activity 2: Identify project risks.
Activity 3: Estimate probability and impact for all risks.
Activity 4: Stratify risks into levels of overall exposure or overall leverage.
Activity 5: Develop strategies to deal with the risks.
Activity 6: Determine the project reserve.
Activity 7: Complete the documentation of the risk management plan.
Activity 8: Get the risk management plan approved.

The activities for dealing with risks during the execution phase are described in Chapter 13, Activity 9.

Purpose of the Risk Management Activities

All projects contain risks. Some risks (threats) may increase costs and project duration, reduce quality, or even jeopardize customer acceptance. Other risks (opportunities) have the potential for reducing costs or project schedule, or increasing revenue, etc. The purpose of the 8 activities is to impose a discipline upon the process, thereby minimizing the negative consequences of threats and increasing the positive effects of the opportunities.

Role of the Project Manager

1. Produce a risk management plan.
2. Monitor risks across the full spectrum of activities.
3. Employ strategies and management behaviors that seek to control risks.
4. Reassess risks periodically.
5. Keep upper management informed.
6. Document the project's risk history: incidences of risk occurrence, actions taken, and final disposition over the full project life cycle; this documentation will contribute to the lessons learned.

Fortunately, the eight activities of project management support these responsibilities.

Outputs of Risk Management

The risk management activities should produce the following desired outcomes:

1. Team members who understand the mission of risk management and their particular roles in this effort
2. A risk management plan that adequately addresses the major threats and opportunities in the project
3. A perception by upper management that the project manager and the team are dealing with project risks successfully
4. A project manager confident in his or her abilities to deal with risks

Risk Management Problems

1. Widespread attitude that a formal risk management process is unnecessary, a waste of time. This is particularly problematic if this attitude exists in upper management. The solution here is to address this with upper management to assure that the project manager is setting a direction that is consistent with what upper management expects. This issue will also require the project manager to set clear expectations of team members because upper managers' attitudes will have permeated to the workers.
2. Team members lacking experience of the project technology may be a source of problems. It will be difficult for a team to develop a risk management plan for a project to develop a software application if none of the members have software application development experience. The solution to this problem is to invite experienced people to join the team during the development of

the risk management plan and to get permanent members with extensive experience in the technology of the project.

3. The eight activities in the process are straightforward, but they do require an understanding of terms and procedures that team members may not have. Training team members in risk management procedures and tools is one way to address this problem. Getting additional members with risk management experience is another.

4. The activities require a discipline that may not be an organizational norm. The project manager may need to demonstrate the desired discipline and to verbally clarify how team members are expected to use the disciplined risk management process.

5. Lack of documentation is a problem more difficult to address. Documentation of what happened on prior projects is crucial to identifying risks, to estimating the probability of risks occurring, and to estimating the impact of these events occurring. If historical data does not exist, there is little the project manager can do to create it. All is not lost but in this circumstance the team is forced to take a subjective approach to assessing risks.

Activity 1: Prepare the Team to Do Risk Management

1. The scope of the risk management effort must be decided. Is the effort intended to cover all phases of the project or is it limited to a particular phase?

2. What information (lessons learned) and documents are available from previous similar projects that will help in our risk assessment?

3. The team will need to gather the work breakdown structure (WBS), network diagram, resource plan, business case definition, and any other documents pertaining to the current project because these documents will be used in our risk assessment.

4. Who are the project's stakeholders and what are their risk-taking propensities and priorities? Is the organization one with strong risk avoidance priorities or is it willing to accept significant risk impacts?

5. Who are the stakeholders and managers who must be kept informed and from whom we will seek advice and acceptance?

6. What is our relationship with these stakeholders? Will they be supportive or critical of our efforts?

7. What is our budget for risk management?

8. What are the formats the team will use to document and communicate risk information?

9. Does the team have the expertise in the technology of the project to manage risks or do we need to seek part-time expertise?

10. Do members of the team have the needed experience in risk management (the procedures and definitions, etc.) or should they be trained?

11. Do team members have the appropriate attitude concerning the discipline required of the risk management process? Does the project manager need to set clear expectations of desired team member behaviors?

12. Each team member will be responsible for monitoring risks in an area of the project, e.g., technical risks, legal risks, procurement risks, resource management risks, etc. The project manager will have to be sure each team member knows his or her area of assigned monitorship and the feedback responsibilities that go along with the assignment. All of the tasks and answers to questions must be documented and retained because they constitute the beginning of the risk management plan.

Activity 2: Identify Project Risks

Risk management includes many terms and concepts that must be defined; these are discussed next.

Risk: A possible event in the future that may have consequences for the project. There are two kinds of risk events: threats and opportunities. A threat is a potential event that would have negative consequences for the project: increased cost, delay in the schedule, reduced quality, or reduced customer satisfaction. An opportunity is a potential event that would have positive consequences for the project: increased revenue, reduced schedules, or enhanced customer satisfaction, etc.

Every risk, whether a threat or an opportunity, has three parts:

1. The risk event itself
2. The probability of the risk occurring
3. The consequence or impact

> **Example 1:** A small consulting company wins a contract to produce a system with which a large government agency can determine its personnel staffing requirements. A handbook on how to conduct personnel staffing studies is the first deliverable for the project. Because the consulting company has little experience with the client, there is the risk that the handbook will not be organized in a way deemed appropriate by the agency. There is a perceived high probability that the client will require extensive rewrites that would cost the company substantial time and money to perform.

> **Example 2:** In the writing of the handbook, the consulting company includes a chapter on advanced regression analysis. Advanced regression analysis includes generating many complex equations with which to estimate the number of people required. The procedure to generate the equations is software-driven and relatively easy. However, interpreting the information to select the best estimating equation requires

an advanced knowledge of regression. The analysts in the government agency don't have the training; this situation presents an opportunity (increased work and revenue) for the consulting company to propose additional work to train the analysts.

Risk statement: Defines the event in a way that clarifies the exact nature of the threat or opportunity. It is difficult to develop strategies for dealing with threats or opportunities if their risk statements are ambiguous or vague. A well-stated risk statement, whether a threat or opportunity, includes

1. What might happen
2. When it might happen
3. The outcome of the event
4. The source of the risk, i.e., the work package (or document paragraph number) to which the risk is attached

Example 3: Source — Submission of deliverable 1 — the handbook: Because we know so little about our customer's organization, there is the threat that the customer may require extensive changes and reorganization of the handbook, thereby causing us to rewrite a large portion.

Output of the Risk Identification Activity

Risk identification will produce two lists: a list of threats on one piece of paper, and a list of opportunities on another piece of paper. Table 9.1 and Table 9.2 show the format for the threat list and the opportunities list, respectively.

Techniques for Identifying Risks

1. **Interview technique:** Gathering threats and opportunities via interviewing requires that the team identify those people who have the experience to help. Interviews may be conducted face-to-face, by telephone, or by instant messaging on the Internet. It's a good idea to have an interview outline if many people are to be interviewed. Table 9.3 is an example of an interview outline.
2. **Records analysis:** Information retained from previous projects can be a rich source of ideas about risks: what threats and opportunities occurred, what action was taken, and what resulted from the action. This kind of information is very useful. After-action reports and accident reports may provide the information from which threats and opportunities can be identified.

   The current project planning documents are a rich source of ideas about threats and opportunities. The WBS is the single best source. The process

Table 9.1   Format for the Threat List

Source	Description of Threat	Probability	Impact	Overall Exposure	Strata

Notes: (1) Use this form to document the threats identified in activity 2 and to document the estimates of probability, impact, and overall exposure determined in activity 3. Activity 4 will document the sorting of these threats into strata by overall exposure. (2) Source is the work package or contract paragraph number to which the threat is connected. The description of the threat must include the specifics of the threat; the description should be a sentence, not a phrase. For probability and impact, indicate the actual numbers or the word scale label from Table 9.4 or Table 9.5. (3) Overall exposure is the product of probability times impact when both probability and impact are expressed as numbers. The product is called the expected value of the threat; i.e., the average cost of the threat if the project were run many times. Use Table 9.7 or Table 9.8 to determine the overall exposure if both probability and impact are expressed in words.

Table 9.2   Format for the Opportunity List

Source	Description of Opportunity	Probability	Impact	Overall Leverage	Strata

Notes: (1) Use this form to document the opportunities identified in activity 2 and to document the estimates of probability, impact, and overall leverage determined in activity 3. Activity 4 will document the sorting of these opportunities into strata by overall leverage. (2) Source is the work package or contract paragraph number to which the opportunity is connected. The description of the opportunity must include the specifics of the opportunity; the description should be a sentence, not a phrase. For probability and impact, indicate the actual numbers or the word scale label from Table 9.4 or Table 9.5. (3) Overall leverage is the product of probability times impact when both probability and impact are expressed as numbers. The product is called the expected value of the leverage; e.g., the average increase in revenue if the project were run many times. Use Table 9.7 or Table 9.8 to determine the overall leverage if both probability and impact are expressed in words.

Table 9.3 Interview Outline

1. Introduce yourself unless the interviewee already knows you.
2. Describe the purpose of the interview and how the information you get will be used.
3. Indicate your appreciation for the time to talk to the interviewee.
4. Indicate that the information the interviewee gives will be confidential. (Sometimes interviewees are hesitant to reveal project experiences because of political reasons.) Confidential means the source (person) of any threats or opportunities will never be disclosed.
5. Ask permission to take notes or use a recorder.
6. Proceed to ask your previously prepared questions. Start out by asking broad, open-ended questions and later asking probing questions. Open-ended questions are non-threatening and help to build rapport because they are easy to answer. Probes increase the tension level because they ask the interviewee to give specifics. Example of open-ended question: Can you suggest any risks that may impact the schedule? Example of probe: You mentioned earlier that upper management attitudes may be a source of schedule threats; can you give me some specifics?
7. When the interview is complete, thank the interviewee, and then ask, "If you were running this project, what threats and opportunities would you focus upon?"

includes reading the work package title; asking the following questions: What could go wrong? What could go well?; and brainstorming answers to these questions.

Example 4: The WBS lists the title of a work package as "Excavate the Basement." (The basement is for a high-rise building in the city where a previous building was demolished.) Threats include the excavation could hit municipal water, electrical, or gas lines, thereby causing a dangerous situation; and the excavation could hit an underground stream or solid rock strata, causing additional cost and schedule delay. The excavation could also uncover an ancient Indian burial ground, thereby creating a political and public relations problem with additional cost and schedule delay or even cancellation of the project.

The network diagram should be examined for schedule threats resulting from dependencies among work packages, work packages with relatively high standard deviations, and near-critical paths, etc.

The contract that initiates the project must also be examined carefully. The procurement decisions concern what services the project will buy from subcontractors and what services it will "buy" from other parts of the organization. The project manager has little recourse when other parts of the organization do not perform as expected. The penalties for nonperformance can be written into an agreement with subcontractors. For this reason, internal "buys" may be more risky than external buys.

3. **Delphi technique:** Delphi is a written survey approach that is used when it is not possible to get the experts to attend a group session because they are geographically dispersed or so busy that clearing a common time and place is almost impossible. An advantage to the Delphi technique is that it gives the respondents the flexibility to provide information at their convenience. Email is the common vehicle. One disadvantage is that respondents may and do procrastinate, so reminders are usually necessary. The process is as follows:

    a. Send out the questions.
    b. Consolidate the responses.
    c. Send the consolidated responses back to responders with a request that they review the consolidated list for additional threats and opportunities.

    The original request for ideas must include a guarantee of confidentiality.

4. **Brainstorming technique:** This is a small-group technique requiring a facilitator/recorder. The advantages of brainstorming are:

    a. It is not threatening.
    b. Participants get a sense of closure from the process; they get a sense of how others react to their ideas.

    The disadvantages include:

    a. It is relatively inefficient. For the time consumed, it does not generate a large number of ideas.
    b. It is subject to manipulation by one person or a small subgroup.
    c. Status differences often affect who speaks and the amount of time that the speaker consumes. Higher-ranking individuals tend to dominate the conversation if they choose to do so.
    d. Some participants are not assertive (because of culture or personality), so their ideas do not get expressed.
    e. It is possible for some people to participate in a brainstorming session and not really say anything. By judicious use of acceptance noises, smiles, and head-nodding an individual may seem to participate, yet contribute nothing. For these reasons, brainstorming requires an experienced facilitator.

    The steps include

    Step 1: The facilitator initiates the discussion by saying, "Our task is to develop a list of threats to the project. Who has an idea?" The facilitator records the ideas as they surface.

    Step 2: When the idea-generation phase is complete, the facilitator leads an effort in consolidating the threats by asking the team to identify any duplications or statements that need to be clarified or reworded.

Step 3: The facilitator asks the group to put the threats into categories. Suggestions might include "technical," "schedule," "manufacturing," and "vendor." The team selects those category names or makes up category names that are appropriate to the project and places the threats under the categories.

The three-step process is repeated for opportunities.

5. **Nominal group technique:** This is a small-group technique also. The advantages are:
   a. It is efficient; it generates a lot of ideas for the time spent.
   b. It is not subject to status differences.
   c. It strongly encourages all members to participate.
   d. It is not subject to manipulation.

   The steps include:

   Step 1: The facilitator initiates the procedure by saying, "Our task is to develop a list of threats to the project. Let's take about 20 minutes for each of you to develop your own list of threats. Record these on a piece of paper. After everyone is done, I'll ask each of you, in turn, to give me one of your ideas. I'll continue going around the table until everyone has exhausted their list." The facilitator may ask the team to focus on a single category of threats, e.g., "let's work on manufacturing risks for the next hour."

   Step 2: The facilitator records all suggestions on large chart paper and consolidates the list to eliminate duplications.

   Step 3: The facilitator has the group put the threats into categories.

   The three-step process is repeated for opportunities.

6. **Crawford slip:** This procedure is similar to the nominal group technique.

   Step 1: The facilitator instructs the group: "Our task is to develop a list of threats to the project. Let's take 20 minutes for each of you to come up with your threat ideas. Record each threat on the self-sticking note sheets. After you are done, stick your ideas on the wall. After everyone is done, we'll eliminate duplications and give each category a name."

   Step 2: The leader facilitates a procedure where the members put the stick-ons into clusters on the wall. They cluster the threats, putting those that seem related together. After they have created the clusters of stick-ons, they give each cluster a category name.

   The same process is used to generate opportunities.

Activity 3: Estimate Probability and Impact for All Risks

This activity consists of two major tasks: estimating the probability of every threat and opportunity that was identified in activity 2, and estimating the impact of every threat and opportunity.

Estimating Probability[4]

There are two approaches to estimating probability of an event happening: the objective numerical approach and the subjective adjective approach. The approach you use depends upon the information you have available. If there is no historical data with which to estimate probability with a number, then you must use a subjective approach. The objective approach estimates probability with a number and the subjective approach estimates probability on a word scale, such as the five-point scale: very low, low, medium, high, and very high probability.

Probability means the likelihood of an event happening. A threat estimated to have a very low probability of occurrence (on a five-point scale) means the team doesn't think it is likely to happen. If the team has historical data relating to the event, it would estimate probability with a number; estimating probability with a calculated number is the preferred approach.

1. **Estimating probability using numbers:** The formula for probability is shown as follows:

   Probability = number of noted events ÷ total number of events (tries)   (9.1)

   Example 5: Of the last 1,000 units manufactured, 15 failed in the first 15,000 hours of service. Our current project will require one of these units to function for 15,000 hours. What is the probability that the unit will fail? The number of noted events, failures, is 15 and the total number of tries is 1,000.

   Probability of failure = 15 ÷ 1,000 = .015 = 1.5%

   Example 6: Over the last three years, our firm has developed five specialized software applications for customers. Of those five efforts, we have failed the customer's total application test three times. We are currently developing a specialized software application and want to estimate the probability of failing the customer's total application test.

   Probability of failure = 3 failures ÷ 5 tries = .60 = 60%

   Probability of passing the customer's test = 2 successes ÷ 5 tries
   = .40 = 40%

Passing or failing the customer's test are mutually exclusive, complementary outcomes of the same risk. There are only two outcomes that could happen; if one outcome occurred (e.g., passed the test), then the other outcome (failed the test) could not have occurred. If you know the probability of one outcome occurring, you can determine the probability of the other by using the one-minus rule: Heads or tails are the two mutually exclusive outcomes of flipping a coin. Mutually exclusive means if you get the head, you cannot get the tail; only one or the other can happen. Equation 9.2 is the one-minus rule, which states that if you know the probability of one of the mutually exclusive events, you can determine the other probability because together they sum to one.

> Probability of one mutually exclusive event
> $= 1 -$ probability of the other mutually exclusive event $\qquad$ (9.2)

Using the one-minus rule, we can estimate the probability of passing the test in example 6 as $1 - .60 = .40 = 40\%$.

Appendix A "Probability Formulas" summarizes six useful probability equations, including Equation 9.1 and Equation 9.2, with examples of how to use each.

2. **Estimating probability using a word scale:** This approach is a viable alternative when there is no historical data with which to estimate probability with a number.

    The subjective approach requires a panel of people who have experience in the technology and categories of risks. For a project which purpose is to develop a specialized software application, the panel needs people knowledgeable of the risks associated with this technology. To estimate risks associated with human resources management, we need someone who has experience in this area. To estimate legal or legislative risk to our project, we need a person experienced in these areas. Invite subject area experts to participate in the risk identification and probability estimating sessions if this expertise does not exist in the team.

    Performing subjective estimates of probabilities involves three steps:

    Step 1: Develop or select a pre-existing word scale. A five-point scale is better than two-, three-, or seven-point scales. Two-point scales (probable versus not probable) are too crude. The three-point scale (low, medium, high probability) is useful but does not provide the number of gradations with which to discriminate among the levels of probability. The five-point scale (very low, low, medium, high, and very high probability) is useful. The seven-point scale requires a higher degree of discrimination among the levels of probability but it can produce improved estimates. The five- and seven-point scales will make activity 4 easier. Table 9.4 and Table 9.5 provide examples of five- and seven-point scales, respectively.

    Step 2: Use a discussion-group consensus or voting process to estimate the probability of each threat and each opportunity. See Table 9.4 and Table 9.5.

    Step 3: Record the estimates of probability for every threat and every opportunity. Use the formats in Table 9.1 and Table 9.2 for threats and opportunities, respectively.

Table 9.4 Example of a Five-Point Probability Scale

Point Labels	Description of Points	Probability Definition
Very low (VL)	Extremely unlikely to happen	10% (one in ten)
Low (L)	Unlikely to happen	25% (one in four)
Medium (M)	Even chances of happening or not happening	50% (one in two)
High (H)	Likely to happen	75% (three in four)
Very high (VH)	Extremely likely to happen	90% (nine in ten)

Notes: (1) The five points for estimating probability are very low (VL), low (L), medium (M), high (H), and very high (VH). (2) Method 1: The team discusses the probability of a threat or opportunity event happening using the description of points and the probability definitions. Once the decision is made, the point label is indicated in the documentation of the List of Threats (Table 9.1) and the List of Opportunities (Table 9.2). Method 2: An alternative method is for the team to discuss and decide the probability, and enter the percent numbers in the List of Threats and List of Opportunities. If Method 2 is used, be sure to indicate that the probabilities are subjective and not the result of calculations.

Table 9.5 Example of a Seven-Point Probability Scale

Point Labels	Description of Points	Probability Definition
Very Very Low (VVL)	Extremely unlikely to happen	10% (one in ten)
Very Low (VL)	Less likely to happen	25% (one in four)
Low (L)	Unlikely to happen	33% (one in three)
Medium (M)	Even chances of happening or not happening	50% (one in two)
High (H)	Likely to happen	66% (two in three)
Very High (VH)	More likely to happen	75% (three in four)
Very Very High (VVH)	Extremely likely to happen	90% (nine in ten)

Notes: (1) The seven points for estimating probability are very very low (VVL), very low (VL), low (L), medium (M), high (H), very high (VH), and very very high (VVH). (2) Method 1: The team discusses the probability of a threat or opportunity event happening using the description of points and the probability definitions. Once the decision is made, the point label is indicated in the documentation of the List of Threats (Table 9.1) and the List of Opportunities (Table 9.2). Method 2: An alternative method is for the team to discuss and decide the probability and enter the percent numbers in the List of Threats and List of Opportunities. If Method 2 is used, be sure to indicate that the probabilities are subjective and not the result of calculations.

Estimating Impact

Impact refers to the consequences of the threat or opportunity happening. These may include increased cost, increased time, reduced quality, loss of equipment, and loss of lives for threats; and increased revenue, reduced cost, reduced schedule, and increased good will for opportunities. As with estimating probability, there is an objective approach, which uses current or historical data, or a subjective approach, which uses a word scale.

1. **Estimating impact with numbers:** There are two approaches to estimating impact with numbers:
   a. The work package approach suggests that the impact of a threat can be estimated in a manner similar to estimating the time and cost to perform a work package. The cost of the threat occurring is the sum of the costs of material, equipment, and labor to deal with the threat event. Threat event impact upon schedule is estimated the same way the duration time for a work package is estimated. Table 9.6 is a suggested

Table 9.6   Risk Cost Estimating Sheet

RISK COST ESTIMATING SHEET							

Project Name: *Estimate by:*   *Approved by:*
Project Manager: *Date:*
Threat/Opportunity Description:

Assumptions/Constraints/Risks:
Labor Cost

Tasks	# People	×	LLR	×	Effort Time	=	Cost
				Total Labor Cost			$__

Materials Cost

Description of Materials	Quantity	×	Unit Cost	=	Item Cost
			Total Materials Cost		$__

Equipment Cost

Description of Equipment	Quantity	×	Unit Cost	=	Item Cost
			Total Equipment Cost		$__
			Total Risk Cost		$__

Problem/Windfall Implementation Time

	Total Implementation Time (Hours)	__

Table 9.6   Risk Cost Estimating Sheet (Continued)

*Instructions: (*1) Use this worksheet to describe the potential risk event (threat or opportunity) and the estimated cost and schedule impacts. (2) Describe the threat or opportunity. (3) Indicate any assumptions or constraints, or risks associated with the potential risk event **Under "Labor Cost":** (4) List the tasks that would have to be performed to deal with the threat or opportunity event. (5) Indicate the number of people who would be needed to recover from the threat or implement the opportunity strategy. (6) Indicate the loaded labor rate (LLR) for each person who would work on recovery/strategy. If two people are to work, indicate their combined LLR. Indicate the combined LLR with the letter "t" after the rate. For example, two people would work on recovery; one has an LLR of $100/hour; the other a rate of $120/hour. Indicate the combined LLR as $220t. (7) Indicate the estimated time to recover/implement under the effort time heading. If the recovery/strategy task is a two-person task and requires that they work together for 40 hours, indicate 40 hours in the "Effort Time" column. Always enter the units of time: hours or days. (8) Multiply # people, LLR, and effort time; enter this figure in the "Cost" column. (9) Sum the figures in the "Cost" column; enter this as total labor cost. **Under "Materials Cost":** (10) List the materials, quantity, unit cost, and item cost for all the materials needed to recover from the threat event or implement the opportunity strategy. (11) Sum the item costs and indicate this figure as total materials cost. **Under "Equipment Cost":** (12) List the equipment, quantity, unit cost, and item cost for all tools and equipment. (13) Sum the item costs and indicate this figure as total equipment cost. **Under "Total Recovery Cost":** (14) Sum the total labor cost, total materials cost, and total equipment cost. Indicate this figure as total recovery cost. **Under "Threat Recovery/Opportunity Strategy Time":** (15) Estimate the time needed to implement the threat recovery or opportunity strategy.

format for a threat-recovery estimating worksheet. This worksheet may be used to document the recovery estimates and actual costs, should the threat occur.

Example 7: A threat exists that the custom software application may fail the total integration test. The software engineering department estimates it will take 160 hours to make this kind of repair. (The estimate must come from people experienced in performing the kind of work required of the threat.) The loaded labor rate of the people who would do the repair is $50/hour. The cost is estimated at $8,000 (160 hours × $50/hour). Two programmers working eight hours a day would need 10 days (160 hours ÷ 16 hours/day) to make the repairs. There are no equipment or material costs. The impact of the threat is estimated at $8,000 in cost and 10 days in schedule time.

b.   The historical data approach uses data from a previous project: The cost of the previous project is adjusted to fit the circumstances of the current threat event.

Example 8: A threat exists that discharge pumps may fail during flooding of low lying areas. The last time we had this problem it cost $7,000 to repair the pumps. Our current deployment of pumps uses twice the number previously used, covering an area twice as large as previously covered. The estimate is the same proportion of pumps will fail and because we have twice as many pumps on line, we estimate the cost at $14,000. The last time we had pump failure it took our crew three weeks to make the repairs; we are estimating six weeks because the possibility exists that twice as many pumps will fail.

Notice that estimating cost and time to recover from threats (or to capture an opportunity) does not involve esoteric math; rather, estimates are made using simple relationships such as there are twice as many pumps now, therefore it will take twice as long to repair the pumps.

2. **Estimating impact with a word scale:** This approach uses word scales to estimate the impact of a threat or opportunity. They are used when it is not possible to calculate impact with numbers. This process requires an estimating team of experienced people who discuss the impact until consensus is reached. Figure 9.1 and Figure 9.2 are Excel® spreadsheets of five- and seven-point subjective impact scales, respectively. Select one of the five- or seven-point scales. Use it to judge the impact on cost and schedule for every threat and opportunity. Record the impacts on the list of threats and the list of opportunities.

Activity 4: Stratify Risks into Levels of Overall Exposure or Overall Leverage

In activity 3, the team estimated the probability and impact for every threat and opportunity. In activity 4, the team will determine the overall exposure for every threat and the overall leverage for every opportunity, and stratify the threats according to overall exposure and stratify the opportunities according to overall leverage. Use Table 9.7 or Table 9.8 to determine the overall exposure of each threat or the overall leverage of each opportunity; enter this information on the threat list or the opportunity list. Table 9.9 and Table 9.10 are examples of threat lists before and after sorting into strata, respectively. Notice that the examples show six threats with estimated probability, impact, and overall exposure. Overall exposure is the average damage that a threat would produce if the project were run many times and the average over the many iterations was calculated. The threats are resorted into "strata order" with the highest exposures on top and threats in the lowest strata on the bottom. Threats are shown in strata order because those in the upper strata must be more closely watched than those in the lower strata. Opportunities must also be shown in "strata order."

Activity 4 is concluded when the team has completed the threat list and an opportunity list, which are similar to the one shown in Table 9.10.

Project Duration = 100 Days

Project Budget = $100,000

	A	B	C	D	E
	Impact Label	**Impact % of Total Schedule**	**Impact in Days**	**Impact % of Total Budget**	**Impact in Dollars**
	Very Low (VL)	Less than: 0.25%	Less than: 0.25	Less than: 0.25%	$250
	Low (L)	0.50%	0.5	0.50%	$500
	Medium (M)	1.00%	1	1.00%	$1,000
	High (H)	5.00%	5	5.00%	$5,000
	Very High (VH)	More than: 5.00%	More than: 5	More than: 5.00%	$5,000

Figure 9.1   Five-Point impact scale.

This impact scale is used when the team needs to select one of the impact points on the five point scale: VL, L, M, H, or VH. Step 1. Enter the project duration in cell E1. Enter the project budget in cell E3. The spreadsheet will change the impact in days (col C) and the impact in dollars (col E) automatically. Step 2. Team estimates the impact in cost of a threat at e.g., $1,000; this translates to an impact label of M. The team estimates that a threat will delay the schedule by more than 5 days; this translates to a VH impact.

This variable five point scale is based on the idea that an impact is deemed high or low etc based on its size relative to the project size. A $1,000 impact on a project with a $2,000 budget is a very high impact whereas a $1,000 impact from a threat on a project with a $200,000 budget is very low. Step 1 above allows the team to enter the total budget and total project duration parameters into the spreadsheet. From these, the spreadsheet calculates the threat costs that correspond to VL, L, M, H, VH impacts. The team estimates (guesses) at the impact in dollars and time and simply reads the impact label corresponding to the judged impact. For example, with a budget of $100,000 and a duration of 100 days, a threat with a $5,000 cost is a high impact and a threat with a one day delay is a medium schedule impact.

| | | Project Duration = | 100 Days | |
| | | Project Budget = | $100,000 | |

A	B	C	D	E
Impact Label	**Impact % of Total Schedule**	**Impact in Days**	**Impact % of Total Budget**	**Impact in Dollars**
Very Very Low (VVL)	Less than: 0.25%	Less than: 0.25	Less than: 0.25%	$250
Very Low (VL)	0.50%	0.5	0.50%	$500
Low (L)	1.00%	1	1.00%	$1,000
Medium (M)	2.00%	2	2.00%	$2,000
High (H)	3.00%	3	3.00%	$3,000
Very High (VH)	5.00%	5	5.00%	$5,000
Very Very High (VVH)	More than: 10.00%	More than: 10	More than: 10.00%	$10,000

Figure 9.2 Seven-Point impact scale.

This impact scale is used when the team needs to select one of the impact points on the seven point scale: VVL, VL, L, M, H, VH or VVH. Step 1. Enter the project duration in cell E1. Enter the project budget in cell E3. The spreadsheet will change the impact in days (col C) and the impact in dollars (col E) automatically. Step 2. Team estimates the impact in cost of a threat at e.g., $1000; this translates to a impact label of Low (L). The team estimates that a threat will delay the schedule by more than 10 days; this translates to a VVH impact.

This variable seven point scale is based on the idea that an impact is deemed high or low etc based on its size relative to the project size. A $1,000 impact on a project with a $2,000 budget is a very high impact whereas a $1,000 impact from a threat on a project with a $200,000 budget is very low. Step 1 above requires the team to enter the total budget and total project duration parameters into the spreadsheet. From these, the spreadsheet calculates the threat costs that correspond to VVL, VL, L, M, H, VH, VVH impacts. The team estimates the impact in dollars and time and simply reads the impact label corresponding to the judged impact. For example, with a budget of $100,000 and a duration of 100 days, a threat with a $5,000 cost is a very high (VH) impact and a threat with a one day delay is a low (L) schedule impact.

Table 9.7 Five-Point Overall Exposure or Leverage Table

Probability	Impact	Overall Exposure or Leverage	Probability	Impact	Overall Exposure or Leverage
Very High	Very High	Very High	Low	Very High	Medium
Very High	High	Very High	Low	High	Medium
Very High	Medium	High	Low	Medium	Low
Very High	Low	Medium	Low	Low	Low
Very High	Very Low	Medium	Low	Very Low	Very Low
High	Very High	Very High	Very Low	Very High	Medium
High	High	High	Very Low	High	Low
High	Medium	Medium	Very Low	Medium	Very Low
High	Low	Medium	Very Low	Low	Very Low
High	Very Low	Low	Very Low	Very Low	Very Low
Medium	Very High	High			
Medium	High	Medium			
Medium	Medium	Medium			
Medium	Low	Low			
Medium	Very Low	Very Low			

Note: Use this table to determine the overall exposure of a threat or the overall leverage of an opportunity if you used the 5 point probability and impact scales. Example: Threat has a probability rating of medium and an impact of low, therefore the overall exposure is low. An opportunity has a probability rating of low and impact of high; the overall leverage is therefore medium.

Table 9.8 Seven-Point Overall Exposure or Leverage Table

Probability	Impact	Overall Exposure or Leverage
Very Very High	Very Very High	Very Very High
Very Very High	Very High	Very Very High
Very Very High	High	Very High
Very Very High	Medium	High
Very Very High	Low	High
Very Very High	Very Low	Medium
Very Very High	Very Very Low	Low
Very High	Very Very High	Very Very High
Very High	Very High	Very High
Very High	High	Very High
Very High	Medium	High
Very High	Low	Medium
Very High	Very Low	Medium
Very High	Very Very Low	Low

(Continued)

Table 9.8   Seven-Point Overall Exposure or Leverage Table (Continued)

Probability	Impact	Overall Exposure or Leverage
High	Very Very High	Very High
High	Very High	High
High	High	High
High	Medium	Medium
High	Low	Medium
High	Very Low	Low
High	Very Very Low	Very Low
Medium	Very Very High	High
Medium	Very High	High
Medium	High	Medium
Medium	Medium	Medium
Medium	Low	Medium
Medium	Very Low	Low
Medium	Very Very Low	Very Low
Low	Very Very High	High
Low	Very High	Medium
Low	High	Medium
Low	Medium	Medium
Low	Low	Low
Low	Very Low	Low
Low	Very Very Low	Very Low
Very Low	Very Very High	Medium
Very Low	Very High	Medium
Very Low	High	Low
Very Low	Medium	Low
Very Low	Low	Low
Very Low	Very Low	Very Low
Very Low	Very Very Low	Very Very Low
Very Very Low	Very Very High	Low
Very Very Low	Very High	Low
Very Very Low	High	Low
Very Very Low	Medium	Very Low
Very Very Low	Low	Very Low
Very Very Low	Very Low	Very Very Low
Very Very Low	Very Very Low	Very Very Low

Notes: (1) Remember the interaction of probability and impact defines leverage for an opportunity. The interaction of probability and impact for a threat defines exposure. (2) Use this table to determine the overall exposure of a threat or the overall leverage of an opportunity if you used the 7 point probability and impact scales. If you used a five-point scale for probability and impact, Table 9.7 is the table you should use to determine overall exposure or leverage. Example: Threat has a probability rating of high and an impact of low; the table indicates this is a medium overall exposure. An opportunity has a probability rating of very low and an impact rating of medium; the table indicates low overall leverage.

Table 9.9   Example of a Threat List BEFORE Sorting by Strata

Source	Description of Threat	Probability	Impact	Overall Exposure
2.1	Basement excavation hits city water line	Medium	$1,000 Medium	Medium
10.5	Electricians strike	High	2-week delay Very High	Very High
8.6	Increase in cost of building materials	.90 High	$20,000 Very High	$18,000 Very High
11.2	Elevator components arrive late	Low	High 1-week delay	Medium
13.4	Dispute with Painters Union delays work	Low	Medium 1-day delay	Low
15.6	Severe weather delays completion of roof	High	High 1-week delay	High

Table 9.10   Example of a Threat List AFTER Sorting by Strata

Source	Description of Threat	Probability	Impact	Overall Exposure or Leverage	Strata
8.6	1. Increase in cost of building materials	.90 High	$20,000 Very High	$18,000	Very High
10.5	2. Electricians strike	High	2-week delay Very High	Very High	Very High
15.6	3. Severe weather delays completion of roof	High	High 1-week delay	High	High
2.1	4. Basement excavation hits city water line	Medium	$1,000 Medium	Medium	Medium
11.2	5. Elevator components arrive late	Low	High 1-week delay	Medium	Medium
13.4	6. Dispute with Painters Union delays work	Low	Medium 1-day delay	Low	Low

Note: Figure 9.1 was used to determine impact ratings. Table 9.7 was used to determine the overall exposure ratings. Table 9.9 is the threat list before sorting.

Activity 5: Develop Strategies to Deal with the Risks

The team will need to develop a strategy for every threat on the threat list and every opportunity on the opportunity list. The only exception to this rule is when the team concludes that a threat has a very, very small exposure that does not require a strategy or an opportunity has a very, very small leverage that does not require a strategy. In activity 5, the team has four tasks to accomplish:

1. Develop strategies for every threat on its threat list
2. Check the interaction effect among threat strategies
3. Develop strategies for every opportunity on the opportunity list
4. Check the interaction effect among opportunities strategies

Develop Strategies for Each Threat

1. **Risk avoidance:** This strategy seeks an alternative method of accomplishing the work package — one that avoids the threat connected to the original method.

   Example 9: The work package title states, "Explore the Bottom of Biscayne Bay." The purpose of the work package is to look for any historical artifacts that may have been deposited on the floor of the bay as a result of the many sailing vessels that have used the bay over the last 500 years. The original method is to walk along the floor of the bay while wearing a diving suit and to explore the surface with a rake. The threat statement is, "There is the threat of being attacked by a shark while walking on the floor of the bay." An avoidance strategy would have to accomplish the same purpose as the original method but completely avoid any chance of a shark attack. Avoidance strategies include using a remote camera, sending divers down in a cage, exploring with a small submarine vehicle, using sonar to detect metal or other large objects, and using a forced-air hose, to blow the sand to uncover artifacts, and a camera, to "see" them.

2. **Accept strategy:** There are two circumstances when it is appropriate or necessary to accept the consequences of a threat event. Sometimes it is appropriate to accept a threat if it has a relatively small cost or schedule impact. Sometimes the project team has no alternative but to accept a threat impact because there is little it can do to alter the probability of it occurring. However, there usually is a lot the project team can do to alter the amount of damage a threat produces by being prepared for it.

   Active acceptance is based on the idea that the damage a threat produces is significantly reduced or contained if the team is prepared for the event before it happens. Active acceptance means having a contingency plan in place before the threat occurs. This means having the people trained with the equipment and materials prepositioned to deal with the threat.

Example 10: The U.S. Navy has been intensely committed to risk management for about 200 years. Here is an example of active acceptance. While in combat, there is the threat that pilots will be wounded when landing on an aircraft carrier. To one side of the landing deck on every carrier is a flight of steps leading to an operating room that is staffed with medical personnel at all times when aircraft are being retrieved. Should a pilot crash on the deck, the pilot would be immediately brought to the operating room for treatment and stabilization until medical evacuation is possible. Having the operating room located close to the flight deck and having it staffed at all times of aircraft retrieval is active acceptance.

A rather unwise alternative to active acceptance is called passive acceptance. Passive acceptance means that the team will do nothing about planning for the threat event but will depend upon its ingenuity after the event occurs. Passive acceptance is the same as crisis management because not preparing for a threat event means you will deal with it as a crisis if and when it occurs. Dealing with threats in the crisis mode greatly increases the cost of recovering because when in crisis, humans are concerned about survival or making the fix and not about cost or efficiency. Sometimes in a crisis our ingenuity fails because it is difficult to be creative when under stress. Sometime ingenuity fails because there is no time to put the ideas into action.

Example 11: While touring the aircraft carrier *U.S.S. Truman,* I noticed a 4″ × 4″ wood beam fastened at the place where the wall (bulkhead) meets the floor (deck). I asked what this was for, and was told, "This passageway is immediately below the flight deck and sometimes an aircraft carrier will take the kind of damage that will cause the flight deck to sag. Planes cannot land on a deck that has a sag in it. The plane will go out of control and destroy the aircraft and possibly kill the crew. If that were to happen, every sailor in this part of the ship knows how to use that beam and the jack positioned a little further down the passageway and others positioned in the ship to jack up the deck." Imagine the panic of trying to fix the deck without the beams and jacks being prepositioned or the sailors trained. Ingenuity alone would not make the fixes quickly enough to prevent aircraft circling above from running out of fuel before they can land.

3. **Risk transfer:** Also called risk deflection. This strategy calls for transferring the consequence of a threat event to another organization. Examples include fire insurance against the threat of a building being destroyed by fire or a contract penalty clause to transfer the consequence of late delivery by the subcontractor. Outsourcing is a threat transfer strategy when the contract specifies that the subcontractor pays the consequences (penalties) for late delivery, unacceptable quality, or cost overrun.

    The project manager cannot transfer the consequences of poor quality or late delivery to another organization when it is the project manager's organization that performs the work that is of poor quality or that arrives late.

4. **Risk mitigation:** This strategy consists of two pre-emptive actions that seek to minimize the probability of the threat occurring or minimize the impact or damage from the threat event.

    Example 11: A threat event for a project to develop specialized software for a customer is described as "Software fails the customer acceptance test." The mitigation strategy is to develop software prototypes for the customer to use. This involves the customer in the software requirements definitions in a way that reduces the likelihood (probability) of failing the customer's acceptance test and also reduces the amount of fixes (impact) that will be required.

Once the strategies have been developed, they are documented on the threat strategies List. See Table 9.11 for an example.

Check Interaction among Threat Strategies

An important step in developing the risk management plan is to assure that strategies employed to ameliorate one threat do not make another threat worse. Table 9.12 describes how to do this. This process may uncover a threat strategy that has negative effects on another of the threats; this strategy will need to be changed. It may uncover a threat strategy that can deal with more than one threat.

Develop Strategies for Each Opportunity[1]

An opportunity is something good that might happen to a project. Examples are opportunities for increased work and revenue or the introduction of a new technology that reduces schedule time. There are four generic strategies for dealing with opportunity events:

1. **Accept the positive impacts (similar to the accept strategy for threats):** There may be active acceptance, where a plan is developed to deal with a potential positive impact before it actually happens. There is passive acceptance, where no action is taken until the event actually occurs. Just as with

Table 9.11 Example of a Threat Strategies List

Threat Number	Threat Strata	Threat Description	Strategy Description
1	VH	Increase in cost of building materials	1A Order materials early 1B Change suppliers to those who won't pass on price increases
2	VH	Electricians strike	2A Negotiate bonus for all electricians if they don't walk off the job 2B Work around the electrical work until strike is over
3	H	Severe weather delays completion of roof	3A Use fabricated tent covers until weather changes
4	M	Basement excavation hits city water line	4A Get city water department to locate the line before we dig
5	M	Elevator components arrive late	5A Establish an early delivery date 5B Accept partial delivery and start work on whatever parts arrive
6	L	Dispute with Painters Union delays work	6A Negotiate bonus for all painters if they don't walk off the job

Instructions for completing the threat strategies list: (1) Give each threat a number starting with the threats in the highest strata and working down to the lowest strata. (2) Indicate the strata level for each threat in the "Threat Strata" column. (3) Indicate the complete description of each threat in the "Threat Description" column. (4) Describe the strategies developed for each threat and give each strategy its unique number. The first strategy for threat 1 is 1A; the second strategy for threat 1 is 1B, etc. The format for the Opportunity Strategies List:

Opport'y Number   Opport'y Strata   Opport'y Description   Strategy Description

Instructions for completing the opportunity strategies list: (1) Give each opportunity a number starting with the opportunities in the highest strata and working down to the lowest strata. (2) Indicate the strata level for each opportunity. (3) Indicate the complete description of each opportunity. (4) Describe the strategies developed for each opportunity and give each strategy its unique number. The first strategy for opportunity 1 is 1A; the second is 1B, etc.

threats, active acceptance has decided advantages: it provides the time to develop an efficient and effective action, whereas passive acceptance relies upon "on-the-spot" ingenuity, which may fail because of lack of planning time or implementation time.

2. **Share the opportunity** means that the project manager will seek to find other parts of the organization that may take advantage of the opportunity because the project manager is not able to do so. Sharing gives up some or

Table 9.12   Example of a Threat Strategy Interaction Analysis

Number/ Strata	Threat Description	Strategy Description	Comments on Interaction Effect among Strategies
1 VH	Increase in cost of building materials	1A Order materials early	Makes 5 worse; not selected
		1B Change suppliers to those that won't pass on price increase	None; 1B selected
2 VH	Electricians strike	2A Negotiate bonus for all electricians if they don't walk off the job	None; 2A selected
		2B Work around the electrical work until strike is over	None; 2B not selected
3 H	Severe weather delays completion of roof	3A Use fabricated tent covers until weather changes	None; selected 3A
4 M	Basement excavation hits city water line	4A Get city water department to locate the line before we dig	None; 4A selected
5 M	Construction materials may be stolen	5A Deploy a six-foot high security fence	None; 5A selected

Instructions for Completing the Threat Strategy Interaction Analysis

1. The interaction analysis form is very similar to the strategies list (Table 9.10). One way to accomplish the analysis is to add the "Comments on Interaction Effect among Strategies" column to Table 9.10. An alternative is to create the strategy interaction analysis form, that is, this table and duplicate the information from Table 9.10.

2. The team reads a strategy and discusses whether it has a positive, negative, or no interaction effect upon the other threats. Does it make a threat worse (negative effect), does it make another threat easier to deal with (positive effect), or does the strategy have no effect upon another threat? After considering these possibilities, the team documents the interaction effect and selects or rejects the strategy.

3. The team continues this process until it has one selected strategy for every threat.

The format for the Opportunity Strategies Interaction Analysis:

Number/ Strata	Opport'y Description	Strategy Description	Comments on Interaction Effect among Strategies

Table 9.12   Example of a Threat Strategy Interaction Analysis (Continued)

Instructions for Completing the Opportunity Strategy Interaction Analysis

1. The interaction analysis form is very similar to the strategies list (Table 9.10). One way to accomplish the analysis is to add the "Comments on Interaction Effect among Strategies" column to Table 9.10. An alternative is to create the form in Table 9.12.
2. The team reads a strategy and discusses whether it has a positive, negative, or no interaction effect upon the other opportunities. After considering these possibilities, the team documents the interaction effect and selects or rejects the strategy. The team documents the decision.

all of the ownership of the opportunity to a third party. This is similar to the transfer strategy for threats.

3. **Ignore the opportunity** means doing nothing before or after the opportunity occurs because the team is not interested in the opportunity. Example: A team is performing a large project that includes some minor software code modifications. The team is able to accomplish these modifications although writing software applications is well outside the organization's expertise. The client signals an opportunity for a follow-up project that would involve the writing of software code for a complicated and complex application. The team's organization decides to ignore (forgo) the opportunity because it lies too far afield from the organization's established expertise and is too risky to undertake.

4. **Enhance the opportunity** has two pre-emptive strategies that either seek to increase the probability of the opportunity being realized or increase the impact/benefit of the opportunity, or both. Enhance strategies, just as mitigation strategies, are pre-emptive in that they are actions taken before the opportunity actually materializes. Enhance strategies are the exact opposite of mitigation strategies. Whereas mitigation strategies seek to decrease the probability and impact of a potential threat, enhance strategies seek to increase the probability and impact of an opportunity.

Check Interaction among Opportunity Strategies

Once the team has developed strategies for the opportunities and documented these strategies on the opportunity strategies list (see Table 9.11), it needs to examine the interaction effect among the strategies using the process described in Table 9.12.

Activity 6: Determine the Project Reserve

A reserve is a set-aside of resources (money, time, people, material, equipment) with which to deal with threat events. Recovering from a threat event will cost time and money; these additional resources should be set aside in advance.

There are two ways to determine the money reserve:

1. **Average cost of all threats:** Requires that the probabilities and impacts of all threats have been estimated with numbers. When probabilities and impacts have been estimated with numbers, it is possible to determine the average cost of each threat. The average cost of a threat over many iterations of the project is called the expected value. The formula is shown in Equation 9.3:

$$\text{Expected value} = \text{probability} \times \text{impact} \qquad (9.3)$$

Example 13: Expected value means average cost of recovering from a threat or the average benefit from an opportunity. Threat A has the following probability and impact, and expected valve

p = 10%	Impact = $10,000	EV = (.10)($10,000) = $1,000

Project Iteration	P	Impact	Actual Occurrence	Actual Cost
1	.10	$10,000	Didn't happen	0
2	.10	$10,000	Didn't happen	0
3	.10	$10,000	Didn't happen	0
4	.10	$10,000	Didn't happen	0
5	.10	$10,000	Didn't happen	0
6	.10	$10,000	It happened	$10,000
7	.10	$10,000	Didn't happen	0
8	.10	$10,000	Didn't happen	0
9	.10	$10,000	Didn't happen	0
10	.10	$10,000	Didn't happen	0

Cost over 10 iterations = $10,000
Average cost = $10,000 ÷ 10 iterations = $1,000.

Example 14: From previous projects, a threat is estimated to have a probability of .25. The impact of the threat is $1,000. There is a 25% chance the threat will occur, and if it occurs it will cost $1,000 to recover. The expected value = $250 (.25 × $1,000 = $250). The average cost of recovering from this threat is estimated at $250.

If we have the average cost of all the threats to the project, we can determine the total average cost of all the threats. If we have the average benefit

Table 9.13   Determining the Reserve

Threat	Probability	Impact (Cost)	Average Cost
Threat #1	.30	$10,000	$3,000
Threat #2	.50	$2,000	$1,000
Threat #3	.25	$6,000	$1,500
Threat #4	.10	$4,000	$400
Total Average Cost			$5,900

Opportunity	Probability	Impact (Additional Revenue)	Average Benefit
Opport'y #1	.25	$4,000	$1,000
Opport'y #2	.3	$6,000	$2,000
Opport'y #3	.1	$3,000	$300
Total Average Benefit			$3,300
Reserve (Method A)			$5,900
Reserve (Method B)			= 2,600

Notes: (1) The project team has identified four threats and three opportunities. The estimates of probability and impact are shown above. The expected value of threat cost or opportunity benefit = probability × impact. (2) Method A: Set the reserve at the average cost of all threats: $5,900. This ignores the benefits. Method B: Set the reserve at the difference between the costs and benefits, i.e., subtract the average benefit of opportunities from the average cost of threats: $2,600 (5,900 – 3,300).

of all opportunities, we can determine the total average benefit from all the opportunities. Combining these will determine the reserve amount. Table 9.13 is an example where a project team has identified four threats and three opportunities.

2. **Historical average of past projects:** The second method for establishing the project money reserve is based on the average cost of threats from previous similar projects.

Example 15: A similar project had a budget of $250,000. When completed, the project had spent $10,000 on recovering from threat events. In the absence of other information, we can set the reserve on the current project at 4% ($10,000 ÷ $250,000 = .04 = 4%). Our project has a budget of $400,000, so we set the cash reserve at $16,000 (.04 × $400,000 = $16,000).

If there are a number of similar projects, the reserve percent can be determined as the average percent. Table 9.14 is an example.

Table 9.14 Determining Average Reserve Percentage Based on Previous Projects

Project	Budget	Cost to Recover from Threats	Percentage
Project A	$1,400,000	$65,000	.0464
Project B	$800,000	$20,000	.0250
Project C	$1,100,000	$45,000	.0409
Project D	$650,000	$22,000	.0338
Project E	$250,000	$6,000	.0240
Total	$4,200,000	$158,000	.0376

The average cost of recovering from threats is 3.76%.

Note: The team can use this average percentage to establish the current project's money reserve. If the total budget for the current project is $500,000, we would set the reserve as $18,800 (.0376 × $500,000 = $18,800). If the team is very conservative, it could use the biggest percentage in the list above (4.64%) and set the reserve at $23,200 (.0464 × $500,000 = $23,200). If the team is very optimistic, it could use the smallest percentage in the list above (2.40%) and set the reserve at $12,000 (.0240 × $500,000 = $12,000). The project team has identified five previous similar projects from which it determines the conservative, average, and optimistic percentage set aside for the money reserve.

There are two ways to determine the time reserve:

1. **Add reserve time to the critical path:** Figure 9.3 shows a network diagram for a project. The duration times were single estimates, i.e., the PERT three-estimate approach was not used. The project duration time is estimated at 40 days, but there is no way to determine the confidence we have in being able to complete the project in 40 days. The critical path goes through work packages A, E, and I. The three-estimate PERT approach was applied to this same project; Table 9.15 shows the optimistic, most likely, and pessimistic estimates for each work package as well as the PERT estimates of average effort time (t_e), standard deviation (σ), and the variance (σ^2).

   Figures 9.3, 9.4, 9.5, and 9.6 provide examples of how to add reserve time to the critical path. Figure 9.3 shows a schedule based on single point estimates. Figure 9.4 shows the same project but where PERT estimates are used. Figure 9.5 shows how to use the PERT information to add risk reserve to the critical path. Figure 9.6 is a spread sheet that can perform the PERT calculations.

2. **Add a reserve of time to the risky work package only:** The second way to build reserve time into the schedule is to select those work packages with a relatively large chance of coming in late and adding a time reserve to each. Work packages with standard deviations approaching 30% of their means are candidates for having their durations expanded to include reserve time. Table 9.15 indicates that work packages B, C, and H have relatively

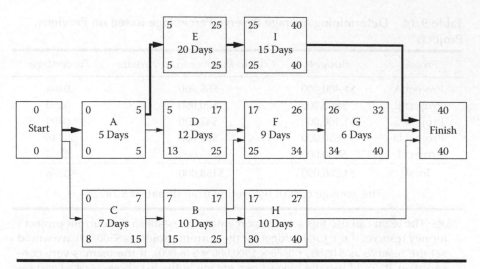

Figure 9.3 Network diagram based on single-point estimates of work package duration time.

Notes:

1. Project duration is estimated at 40 days.
2. Single point estimates provide no information with which to estimate the standard deviations of work package durations. Without this, we can not identify those work packages that may take significantly longer to complete than the estimate. (See Table 9.15 Note 2.)
3. Without standard deviations, we can not determine the amount of reserve time to add for estimating risk.
4. Without the standard deviation of the critical path, we have no way of estimating our confidence in completing the project in 40 days.
5. Table 9.15 and Figure 9.6 which demonstrate the PERT three point approach.
6. For alternative methods of estimating the project's duration, see the article, "Adding Risk into Project Estimates", Endnote 2.

large standard deviations. A large standard deviation means there is a good chance that the actual duration may be significantly greater or less than the estimated mean. If we add 1.28 standard deviations to each of these risky work packages, we determine a new duration in which we can be 90% confident; see Equation 9.6. If we add two standard deviations to each of these risky work packages, we determine a new duration for each in which we can be 95% confident; see Equation 9.5. The matrix in Table 9.15 shows a simple way to increase our confidence.

Step 1: Square the standard deviations of all the work packages on the critical path; the squared standard deviations are called variances. Table 9.15 shows the variances for the work packages (A, E, and I on the critical path) as .69, 6.25, and 1.77 respectively.

Table 9.15   PERT Data Matrix

Work Package	Optimistic	Most Likely	Pessimistic	Average t_e	Std Dev σ	Variance σ^2	Critical Path Variance σ^2
A	3	5	8	5.2	.83	.69	.69
B	5	10	30	12.5	4.17	17.39	
C	4	7	24	9.3	3.33	11.09	
D	8	12	18	12.3	1.67	2.79	
E	15	20	30	20.8	2.5	6.25	6.25
F	7	9	12	9.2	.83	.69	
G	4	6	9	6.2	.83	.69	
H	7	10	40	14.5	5.50	30.25	
I	12	15	20	15.3	1.33	1.77	1.77
			Sum of the variances on the critical path =				8.71

Standard deviation of the critical path = σ_{cp} = square root of 8.71 = 2.95 days

Figure 9.6 is an Excel® spreadsheet which performs the calculations in this matrix.

Notes: (1) This chart shows the PERT information for all the work packages in Figure 9.4; the average time (t_e) was used as the duration for the work packages. (2) Work packages B, C, and H have a strong likelihood of completing very late because their standard deviations are relatively large; approximating 30% of the work package mean. (3) The variances for those work packages on the critical path (A, E, and I) were summed. The square root of the summed variances equals the standard deviation of the critical path. (4) PERT formulas:

Work package average duration (t_e) = [Pessimistic + (4 times Most Likely) + Optimistic] ÷ 6

$$t_e = [P + 4(ML) + Op] \div 6 \qquad (5.1)$$

Work package standard deviation = σ = (Pessimistic − Optimistic) ÷ 6

$$\sigma = (P - Op) \div 6 \qquad (5.2)$$

Standard deviation of critical path = square root of sum of variances (i.e., squared standard deviations) of work packages on the critical path

$$\sigma_{cp} = \sqrt{(\sigma^2_A + \sigma^2_E + \sigma^2_I)} \qquad (9.4)$$

Max time to complete the project at 95% confidence equals the average time to complete it plus 2 critical path standard deviations

$$\text{Max time at 95\% confidence} = \text{path } t_e + 2\sigma_{cp} \qquad (9.5)$$

Max time to complete the project at 90% confidence equals the average time to complete it plus 1.28 times the critical path standard deviation

$$\text{Max time at 90\% confidence} = \text{path } t_e + 1.28\sigma_{cp} \qquad (9.6)$$

Equation 9.5 is used in Figure 9.5 to add six days of reserve time to the schedule.

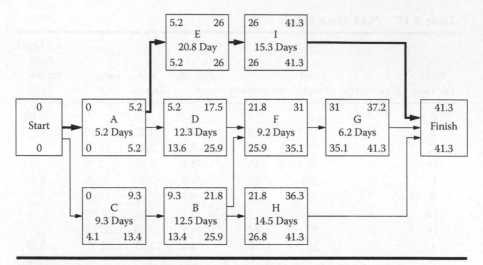

Figure 9.4 Network diagram based upon PERT estimates of work package duration time.

Notes:

1. Average project duration is estimated at 41.3 days. Because the project duration time is an average, we can only be 50% confident that we can complete the project in 41.3 days.
2. The average duration (t_e) and the standard deviation (σ) for each work package are shown in Table 9.15.
3. Table 9.15 determined the critical path standard deviation. With this, we may add a time reserve to the project duration time to determine a project duration in which we are 95% confident. Figure 9.5 shows how this is done.

Step 2: Sum of variances equals 8.71.

Step 3: Take the square root of the sum. The square root of the sum of the variances is called the standard deviation of the critical path (σ_{cp}). The σ_{cp} equals 2.95 rounded to 3 days.

Step 4: Add two critical path standard deviations (six days) from Table 9.15 to the average project duration time (41.3 days) to determine the project duration of 47.3 days; this is Equation 9.5 from Table 9.15. We are 95% confident that we can complete the project in 47.3 days. The six days is added to the network diagram as 6R, indicating that it is 6 days of reserve time. Figure 9.5 shows how this reserve time is determined and how it is indicated on the network diagram. Figure 9.6 is an Excel® spreadsheet that performs the PERT calculations and determines the amount of reserve time to add to achieve confidence at the 90% and 95% levels.

All projects should have estimated project duration times in which the project manager is 90% or 95% confident. The spreadsheet in Figure 9.6 makes this easy to achieve.

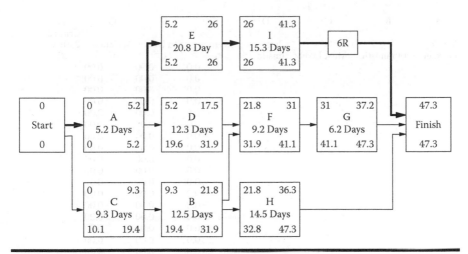

Figure 9.5   Network diagram with six days reserve time.

Notes:

1. Table 9.15 shows the standard deviation of the critical path is 2.95 days. Figure 9.4 shows the average project duration is 41.3 days. This means we are 50% confident that the project can be completed in 41.3 days.
2. The project duration at 95% confidence, (see Equation 9.5 on Table 9.15) = the project average duration plus two critical path standard deviations. Two standard deviations equal 6 days (2 × 2.95 = 5.90 rounded to 6).
3. The project duration above is changed from 41.3 to 47.3 days (this includes the 6 days reserve).
4. The six days of reserve time are indicated on the critical path arrow immediately preceding the finish box. The symbol 6R on the critical path indicates that the project has 6 days of reserve time.
5. Complete the forward pass being sure to include the 6 days of reserve time. Recalculate the backward pass because the late schedule for all work packages not on the critical path will change. For purposes of the forward and backward passes, the 6R is treated as if it were a lag. On the forward pass, add the 6 and on the backward pass, subtract the 6. The "R" is intended to indicate that the six is a reserve and not a lag.
6. Note that the result of adding six days reserve time to the critical path has increased the total float of all work packages not on the critical path by six days. This makes the project much easier to achieve.
7. The Excel® spreadsheet in Figure 9.6 calculates (a) the standard deviation of the critical path, (b) the reserve time at 95% confidence and (c) the reserve time at 90% confidence. Choose (b) or (c) and include the reserve time into the network diagram in accordance with steps 4 and 5 above.

A	B	C	D	E	F	G	H
							Critical
Work			Pessi-	Average	Std Dev	Variance	Path Var.
Package	Optimistic	Most Likely	mistic	t_e	σ	σ^2	σ^2
				0.0	0.00	0.00	
				0.0	0.00	0.00	
				0.0	0.00	0.00	
				0.0	0.00	0.00	
				0.0	0.00	0.00	
				0.0	0.00	0.00	
				0.0	0.00	0.00	
				0.0	0.00	0.00	
				0.0	0.00	0.00	
				0.0	0.00	0.00	
				0.0	0.00	0.00	
				0.0	0.00	0.00	
				0.0	0.00	0.00	
				0.0	0.00	0.00	
				0.0	0.00	0.00	
				0.0	0.00	0.00	
				0.0	0.00	0.00	
				0.0	0.00	0.00	
				0.0	0.00	0.00	
				0.0	0.00	0.00	
				0.0	0.00	0.00	
				0.0	0.00	0.00	
				0.0	0.00	0.00	
				0.0	0.00	0.00	
				0.0	0.00	0.00	
				0.0	0.00	0.00	
				0.0	0.00	0.00	
				0.0	0.00	0.00	
				0.0	0.00	0.00	

Sum of the variances on the critical path =   0.00 Days

Square root of the Sum of variances on the critical path =   0.00 Days

Standard deviation of the critical path =   0.00 Days

Reserve time for 95% confidence =   0.00 Days

Reserve time for 90% confidence =   0.00 Days

Figure 9.6   Excel® spreadsheet that calculates the time-reserve for the critical path.

Notes:

1. Enter the work package numbers in column A.
2. Enter the optimistic estimates of work package durations in column B.
3. Enter the most likely estimates of work package durations in column C.
4. Enter the pessimistic estimates of work package durations in column D.
5. The spreadsheet will calculate the weighted average duration for each work package and display it in column E.
6. The spreadsheet will calculate the work package standard deviation and display it in column F.

7. The spreadsheet will calculate the variance for each work package and display it in column G.
8. Use the average durations displayed in column E to create the network diagram. Determine the work packages that are on the critical path. Figure 9.4 indicates A, E and I as critical.
9. Copy the variance for each critical work package from column G to column H.
10. The spreadsheet will calculate the standard deviation for the critical path and display it in H38.
11. The spreadsheet will calculate the reserve time that must be added to the project network diagram to set a total duration that we are 95% confident we can achieve. This reserve is displayed in H40.
12. The spreadsheet will calculate the reserve time that must be added to the project network diagram to set a total duration that we are 90% confident we can achieve. This reserve is displayed in H42.
13. Refer to Figure 9.5 for instructions on how to include reserve time in the network diagram.

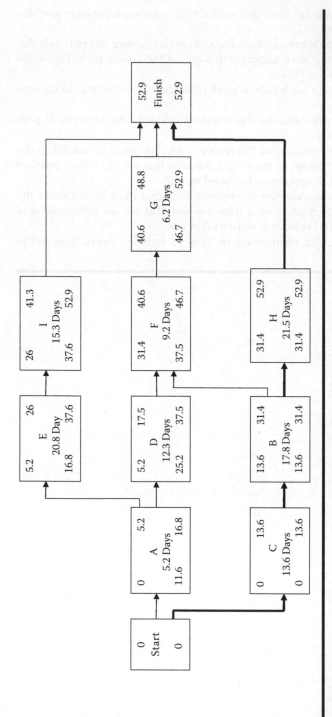

Figure 9.7 Network Diagram with reserve-time added to risky work packages.
The durations of B, C and H have been increased by 1.28 standard deviations to indicate durations in which we are 90% confident.

2. Add a reserve of time to the risky work packages only.

The second way to build reserve time into the schedule is to select those work packages with a relatively large chance of coming in late and adding a time reserve to each. Work packages with standard deviations approaching 30% of their means are candidates for having their durations expanded to include reserve time. Table 9.15 indicates that work packages B, C, and H have relatively large standard deviations. A large standard deviation means there is a good chance that the actual duration may be significantly greater or less than the estimated mean. If we add 1.28 standard deviations to each of these risky work packages, we determine a new duration in which we can be 90% confident; see Equation 9.14. If we add two standard deviations to each of these risky work packages, we determine a new duration for each in which we can be 95% confident; see Equation 9.13.

Work package B has a mean of 12.5 and a standard deviation of 4.17. This work package is risky because its standard deviation is 33.4% (4.17 ÷ 12.5) of the mean. The duration at 90% confidence equations 17.8 (12.5 + [1.28 × 4.17])

Work package C has a mean of 9.3 and a standard deviation of 3.33. The duration at 90% confidence is 13.6 (9.3 + [1.28 × 3.33]).

Work package H has a mean of 14.5 and a standard deviation of 5.50. The duration at 90 confidence is 21.5 (14.5 + [1.28 × 5.50])

Using the B, C and H durations of 17.8, 13.6 and 21.5 in the network diagram instead of the original durations will increase the project duration from 41.3 (Figure 9.4) to 52.8 (Figure 9.7).

Figure 9.7 shows the impact of adding time-reserve to work packages B, C and H:

1. Adding time-reserve to risky work packages (method 2) may and often determines a total project duration that is greater than that produced by adding a time-reserve to the critical path (method 1). Method 1 determined a project duration of 47.3 days (Figure 9.5) and method 2 produced a project duration of 52.9 (Figure 9.7). Method 2 is the more conservative of the two methods.

2. The two methods may project two different critical paths. In Figure 9.5 (method 1), the critical path goes through work packages A, E and I. In Figure 9.7 (method 2), the critical path goes through work packages B, C and H.

3. The total float on the noncritical work packages is larger with method 2 than it is with method 1. Since total float is a reserve of time for the noncritical work packages, method 2 provides more reserve time.

Which to use? Adding time-reserve to the critical path makes intuitive sense since it is the critical path that determines the project duration. However method 1 determines the critical path without knowledge of the risky work packages. Method 2 determines the risky work packages and uses the expanded durations of these to determine the critical path because it may be the risky work packages that determine the project duration.

Activity 7: Complete the Risk Management Plan Documentation

This activity consists of one major task: assemble all the documentation generated from activities 1 through 7 into a risk management plan document.

Activity 8: Get the Risk Management Plan Approved

The risk management plan must be briefed and staffed to the project stakeholders and upper management to solicit their advice and consent, and to assure that upper management is informed of the decisions the team has made in relation to risk assessment, impacts, strategies, etc. The project manager's strategies and actions must be consistent with the priorities of upper management. This applies to the entire project but especially to risk management.

Notes

1. *Project Management Body of Knowledge* (PMBOK®), Third Edition, 2004. Project Management Institute, Four Campus Boulevard, Newtown, PA 19073. Page 261 includes exploit as an opportunity strategy, whereas it has been omitted from this book because the exploit strategy is similar to the enhance strategy. Enhance and exploit differ only in the degree of probability they seek. Enhance seeks to increase probability whereas exploit seeks to achieve 100% probability. Any strategy that is called an exploit strategy can be correctly classified as an enhance strategy. This book adds a fourth strategy for opportunities, the ignore strategy, because there are times when an organization chooses not to pursue an opportunity.
2. "Adding Risk into Project Estimates — PERT vs. Monte Carlo," by Guy L. De Furia, Ph.D., PMP. (Article first appeared in the April 2007 issue of *ESI Horizons*. © ESI International, 901 North Glebe Road, Suite 200, Arlington, VA 22203; www.esi-intl.com. With permission.)
3. "How to Estimate Risk Probabilities," by Guy L. De Furia, Ph.D., PMP. (Article first appeared in the November 2007 issue of *ESI Horizons*. © ESI International. 901 North Glebe Road, Suite 200, Arlington, VA 22203; www.esi-intl.com. With permission.)
4. Project managers are not expected to be experts in all disciplines. If you are confronted with a probability problem that you are unfamiliar with, contact an experienced statistician, as did this author. The author thanks Ms. Mary Erickson of the Department of Commerce, National Ocean and Atmospheric Administration, for her help in clarifying the procedures used to determine risk probabilities. Errors, if any, rest with the author.

Appendix A

Probability Formulas[4]

Estimating the probability of an event requires that you first create a fraction.

$$\text{Probability} = \text{number of noted events} \div \text{total number of events} \quad (9.1)$$

Example 1: In the last five years we have developed application software for eight different clients. In six of these projects, we have failed the customer initial system integration test. What is the probability of failing this test on the current project?

The noted event is failing the test. Number of noted events is six; the total number of events (projects) is eight.

$$\text{Probability} = 6 \div 8 = .75 = 75\%$$

Mutually exclusive events: Mutually exclusive events are the possible outcomes of a single risk where the occurrence of one possibility automatically precludes the occurrence of the complementary event. In flipping a coin, getting a "head" or getting a "tail" are complementary events. If we get the head it means we did not get and cannot get the complementary event — the tail.

If you know the probability of one mutually exclusive event, you can easily estimate the probability of its complement by using the one-minus rule (formula): If A and B are mutually exclusive complementary events from the same risk, then

$$\text{Probability of } B = 1 - \text{probability of A}$$

$$\text{Probability of } A = 1 \text{ probability of B} \quad (9.2)$$

Example 2: If the probability of passing a test is 80%, then the probability of not passing is $1 - .80 = .20 = 20\%$.

The probability of event A occurring or its complement (event B) occurring is equal to the sum of their individual probabilities.

$$P(A) \text{ or } P(B) = P(A) + P(B) \tag{9.7}$$

Please note that Equation 9.7 applies to the mutually exclusive (complementary) events from the same risk, not from two different risks.

Example 3: If the probability of getting a flat tire is .10 and the probability of its complement, not getting a flat tire, is .90, then the probability of getting or not getting a flat is obviously .10 + .90, which is equal to 1.00. But if the probability of getting a flat tire is .10 and the probability of running out of gas is .30, the probability of getting a flat tire or running out of gas is not .10 + .30 = .40 because getting a flat tire and running out of gas are not two possible outcomes of the same risk. They are two different outcomes from two different and independent risks!

Independent events are events where the occurrence of one event has no effect upon the probability of an event from another and different risk.

Example 4: Two people are drawing playing cards at different tables. John is trying to draw a "spade" from the deck and Mary is trying to draw a "king." What happens at one table has no effect upon what happens at the other table. The events are independent of each other.

From Equation 9.1 we estimate the probability of John drawing a spade (event S) from his deck is $P(S) = 13/52 = .25$ because there are 13 spades (the noted event) in the deck of 52 cards. The probability of Mary drawing a king (event K) from her deck is $P(K) = 4/52 = .08$ because there are four kings (the noted event) in the deck of 52 cards.

What is the probability that both John and Mary will be successful; that John draws a spade and Mary draws a king?

The probability that two or more independent events will occur is equal to the product of their individual probabilities.

$$P(S) \text{ and } P(K) \text{ both happening} = P(S) \times P(K) \tag{9.8}$$

The probability of both happening is .25 x .08 = .02 = 2%.

Example 5: What is the probability that either John or Mary will be successful, but not both?

$$P(S) + P(K) - 2P(S \text{ and } K) \tag{9.9}$$

Events S and K are independent events. P(S and K) means probability of getting the king of spades = 1/52 = .02.

For example 5, P(S) = .25; P(K) = .08, P(S and K) = .02; therefore the probability of S or K occurring but not both = .25 + .08 - 2(.02) = .29 = 29%.

Sometimes we are interested in predicting the probability of one or another of two independent events occurring or both occurring.

$$P(H) + P(V) - P(H \text{ and } V) \tag{9.10}$$

Example 6: We have a long flight over open water. There are two risks: (1) running low on fuel because we are flying against strong headwinds and (2) we encounter violent weather. From previous flights we determine that the probability of encountering strong headwinds is 60% and the probability of violent weather is 40%. What is the probability that we will experience strong headwinds or violent weather or both? The strong headwinds and violent weather are independent risks and they could both occur. We define headwinds as event H and violent weather as event V.

P(H) = .60, P(V) = .40, P(H and V both occurring)
= .60 × .40 = .24

P(H) + P(V) - P(H and V) = .60 + .40 - .24 = .76 = 76%

We have a 76% chance that we will encounter one or both of these problems in our next flight.

What is the probability of encountering one or the other of these problems but not both? Equation 9.5 applies. Probability of H or V but not both = P(H) + P(V) - 2P(H and V) = .60 + .40 - 2(.60 × .40) = .52 = 52%.

Chapter 10

Project Baseline and Control[1]

Mega Recipes for Determining the Baseline and Control Limits

Activity 1: Establish the cost and schedule baselines.
Activity 2: Establish the terminal control limits.
Activity 3: Set instrumental control limits.
Activity 4: Communicate the instrumental control limits.
Activity 5: Monitor performance against the control limits.
Activity 6: Evaluate project performance.
Activity 7: Determine critical path SPI.
Activity 8: Forecast project final duration.
Activity 9: Forecast project final cost.

Role of the Project Manager

The project manager's primary role during the project execution phase, assuming that the cost and duration of each work package have been properly estimated, is to establish and communicate the instrumental control limits for each work package. In addition, the project manager must monitor work package performance against these limits. The project manager will also need to periodically evaluate the total project performance and forecast final project cost and duration.

Purpose of Control

The purpose of all control is to keep variance within acceptable limits. Variance is defined as the difference between what is planned and what is actually achieved

$$\text{Variance} = \text{plan} - \text{actual} \qquad (10.1)$$

Example 1: A project with a planned budget of $100,000 is accomplished at an actual cost of $120,000. The cost variance is -$20,000 ($100,000 - $120,000). The project is $20,000 over the original budget. If being $20,000 over budget is acceptable to the customer, then the project manager has no problem. If being $20,000 over budget is more than what is acceptable, then the project manager should have controlled the project to keep the cost within acceptable variance. If being $20,000 over budget is acceptable variance to the customer, then the cost is said to be "in control." If $20,000 is more than what is acceptable to the customer, then the cost of the project is said to be "out of control."

Example 2: A project has a scheduled duration time of 80 days. The project is actually completed in 95 days. Schedule variance is -15 days (80 days - 95 days), indicating 15 days late. If being three weeks late is more than what is acceptable, then the project manager has not controlled the schedule adequately. Depending upon what the customer considers acceptable schedule variance, this project may be "in control" or "out of control" for schedule.

Control Defined

The concept of project control has been borrowed from the physical sciences. To an engineer, a control is a device that does three things:

1. Establishes a performance standard.
2. Continuously monitors performance.
3. Intervenes when performance deviates too much from what is desired.

Example 3: The thermostat in a room is an example. We set the desired temperature, e.g., 74 degrees. The thermostat monitors the temperature. When the thermostat senses that the temperature is 70 degrees, it sends an electrical signal to the furnace to send more heat to the room. When the temperature of the room reaches 76 degrees, the thermostat sends another signal to the furnace — this time to stop sending heat. The desired temperature is 74 degrees; the upper control threshold is 76 degrees, and the lower control threshold is 70 degrees. The performance standard is 74 degrees; 76 is the upper control limit, and 70 is the lower control limit. As long as the temperature is more than 70 but less than 76 degrees, the control device takes no action. The control device intervenes when the temperature is either above the upper limit or below the lower limit.[2]

In project management, the desired performance level is called the baseline (e.g., 74 degrees). It is the project manager's job to keep performance within the upper (76 degrees) and lower (70 degrees control limits). For each work package, there will be a baseline with upper and lower limits for costs as well as a baseline with upper and lower limits for schedule. For project management, the instrumental baselines are developed from the terminal baselines.

Baseline Defined

The baseline is the customer's expectations relative to scope, cost, and time. The baseline indicates what the customer will get (i.e., deliverables), how much the project will cost, and how long the project will take to be completed. The customer's expectations relative to the final cost and duration of the project are terminal baselines because they reflect the expectations for the total project — the end of the project. To accomplish the terminal baselines, the project manager establishes instrumental baselines for each work package with the strategy that achieving these instrumental baselines will produce the project's terminal expectations (baselines).

Keeping each work package within its instrumental expectations will achieve the terminal expectations. This is analogous to Ben Franklin's suggestion: "Watch the pennies and the dollars will take care of themselves."

Instrumental baselines are established for each work package. The budget for each work package is the cost baseline for that work package. The duration estimate is the duration baseline. (The work package scope baseline is documented in the work package work order.) The project manager seeks to accomplish each work package at a cost close to the cost baseline (within acceptable cost variance) and with a duration approximating the duration baseline (within acceptable schedule variance). To assure that each work package variance is within acceptable limits requires activities 1 through 5.

Activity 1: Establish Cost and Schedule Baselines

1. Establish the budget for each work package. (See Chapter 5.) These estimates, once approved, become the cost baselines for each work package.
2. Establish the total project budget. (See "Bottom-Up Estimate of Project Cost" in Chapter 5.) The budget for the project will become the cost baseline for the project.

3. Estimate the duration time for each work package. (See "How to Estimate Work Package Duration Times" in Chapter 6.) This estimate will become the duration baseline for each work package.
4. Determine the project total duration time. (See activity 2 in Chapter 6.)

Example 4: A project manager is assigned to manage project A. The project manager developed the work breakdown structure and estimated the cost of each work package. From these, she estimated the total project cost at $500,000. She estimated the duration time of each work package and developed the network diagram; this showed that the project had a duration time of 100 days. The cost and schedule terminal baselines are $500,000 and 100 days, respectively. The budget and duration estimates for the work packages are the instrumental cost and schedule instrumental baselines, respectively.

Activity 2: Establish Terminal Control Limits

The limits of acceptable variance around the cost performance index (CPI) will be used to determine if actual cost is within acceptable variance. The limits of acceptable variance around the schedule performance index (SPI) will be used to determine if performance against the schedule is acceptable (see activity 6 in Chapter 13):

1. Determine the maximum and minimum acceptable project costs. See example 5.
2. Determine the maximum and minimum acceptable project duration time. See example 5.

Example 5: The project manager in example 4 knows that completing a project to the exact budgeted cost and duration time are unrealistic goals. She negotiates with the sponsor, who decides that it would be acceptable if the total project cost were as high as $525,000 or as low as $425,000.[3] The sponsor requires that the project be completed in 115 days maximum and 85 days minimum.

Maximum acceptable project cost	=	$525,000
Baseline project cost	=	$500,000
Minimum acceptable project cost	=	$425,000
Maximum acceptable project duration	=	115 days
Baseline project duration	=	100 days
Minimum acceptable project duration	=	85 days

If the sponsor expresses no interest in discussing maximum and minimum cost or duration times, the project manager should establish these for purposes of controlling the project. If necessary, these need not be disclosed to the sponsor or customer because they are instrumental guide markers to help the project manager achieve the sponsor's cost and duration expectations.

3.   Determine the lower control limit (LCL_{cpi}) for cost[4]:

$$LCL_{cpi} = \text{cost baseline} \div \text{maximum acceptable cost} \qquad (10.2)$$
$$= \$500,000 \div \$525,000 = .952$$

4.   Determine the upper control limit (UCL_{cpi}) for cost:

$$UCL_{cpi} = \text{cost baseline} \div \text{minimum acceptable cost} \qquad (10.3)$$
$$= \$500,000 \div \$425,000 = 1.176$$

For this project, the strategy for controlling project costs is to keep the CPI for every work package between the limits of .952 and 1.176. The project's maximum acceptable cost of $525,000 will not be exceeded if the cumulative CPI for the total project is .952 or larger. The project's total cost will not be less than the minimum acceptable cost of $425,000 if the cumulative CPI for the total project is 1.176 or less (but not less than .952).

5.   Determine the lower control limit LCL_{spi} for schedule[5]:

$$LCL_{spi} = \text{duration baseline} \div \text{maximum acceptable duration} \qquad (10.4)$$
$$= 100 \text{ days} \div 115 \text{ days} = .870$$

6.   Determine the upper control limit UCL_{spi} for schedule:

$$UCL_{spi} = \text{duration baseline} \div \text{minimum acceptable duration} \qquad (10.5)$$
$$= 100 \text{ days} \div 85 \text{ days} = 1.176$$

The strategy for controlling this project's schedule is to keep the SPI for every work package between the limits of .870 and 1.176. The project's maximum acceptable duration will not be exceeded if the cumulative SPI for the total project is .870 or larger (but not more than 1.176). The project's total duration will not be less than the minimum acceptable duration if the cumulative SPI for the total project is 1.176 or less (but not less than .870).

Total project CPI and SPI thresholds can and should be established immediately after the sponsor or the customer has approved the project budget and duration baselines.

Work package managers need to know the maximum and minimum acceptable costs for their work packages. They also need to know the maximum and minimum duration times for their work packages. Therefore the project manager must set instrumental control limits for each work package.

Activity 3: Set Instrumental Control Limits for Each Work Package

Example 6: Work package A in project A (see example 4) has a cost baseline (budget) of $10,000. The work package duration baseline is five days. The project manager knows the upper and lower cost control limits are 1.176 and .952, respectively. The upper and lower duration control limits are 1.176 and .870, respectively. From these, the maximum and minimum acceptable costs and durations are determined.

1. Set cost limits for work package (WP) A:

$$\text{Maximum acceptable cost} = \text{WP cost baseline} \div \text{lower control}$$
$$\text{limit for cost} = \$10,000 \div .952 = \$10,504 \qquad (10.6)$$

$$\text{Minimum acceptable cost} = \text{WP cost baseline} \div \text{upper control}$$
$$\text{limit for cost} = \$10,000 \div 1.176 = \$8,503 \qquad (10.7)$$

2. Set duration limits for work package A:

$$\text{Maximum acceptable duration} = \text{WP duration baseline} \div \text{lower}$$
$$\text{control limit for schedule} = 5 \text{ days} \div .870 = 5.75 \text{ days} \qquad (10.8)$$

$$\text{Minimum acceptable duration} = \text{WP duration baseline} \div \text{upper}$$
$$\text{control limit for schedule} = 5 \text{ days} \div 1.176 = 4.25 \text{ days} \qquad (10.9)$$

Activity 4: Communicate the Instrumental Baseline and Control Limits

Communicate the maximum and minimum acceptable work package costs and the maximum and minimum acceptable duration times to the person responsible for work package A. Document these limits on the work package work order.

Table 10.1 is an example of the variation control chart. This chart summarizes the acceptable cost and schedule variances for every work package. Table 10.2 is a template for building this chart. Figure 10.1 is an example of a spreadsheet-generated control chart. Figure 10.2 is a spreadsheet template that can perform all

Table 10.1   Variation Control Matrix for Example 4 and Example 5

Work Package	Cost Baseline	Minimum Accept. Cost	Maximum Accept. Cost	Duration Baseline	Minimum Accept. Duration	Maximum Accept. Duration
A	$10,000	$8,503	$10,504	5 days	4.25 days	5.75 days
B	$20,000	$17,007	$21,008	10 days	8.50 days	11.50 days
C	$5,000	$4,252	$5,252	5 days	4.25 days	5.75 days
D	$25,000	$21,258	$26,260	13 days	11 days	15 days
E	$30,000	$25,510	$31,513	15 days	12.75 days	17.25 days
Total Project	$500,000	$425,170	$525,210	100 days	85 days	115 days

Note: This table shows the baseline cost, minimum acceptable cost, and maximum acceptable cost for each work package plus the baseline, minimum, and maximum acceptable costs for the total project. It also shows the baseline duration, minimum acceptable duration, maximum acceptable duration for every work package, and the baseline duration, and minimum and maximum acceptable duration times for the total project. Not all work packages are shown.

the calculations. The project manager uses the variation control chart to establish the limits of acceptable cost and schedule on every work package. This information is indicated on the work package work order.

Activity 5: Monitor Performance against the Control Limits

The project manager's purpose in establishing control limits is to keep the total project cost and schedule variance within acceptable limits. The project manager may use the variation control matrix to monitor individual work package cost and schedule performance as well as monitor total project performance.

With these prerequisites in place, the project manager is in a position to monitor each work package and evaluate project performance.

Activity 6: Evaluate Project Performance

Example 7: Project A in example 4 is at day 25 in its schedule. Table 10.3 shows the information we have: Work package A is $1,000 worth of work behind schedule (SV = –$1,000). The SPI index of .900 indicates that being $1,000 worth of work behind schedule is acceptable because .900 falls between the upper and lower SPI control limits of .870 to 1.176. Work package B and work package C are over budget by $2,000 and

Table 10.2 Manual Template for a Variation Control Matrix

Work Package	Cost Baseline	Minimum Accept. Cost	Maximum Accept. Cost	Duration Baseline	Minimum Accept. Duration	Maximum Accept. Duration
Note 1	$(Note 2)	$(Note 3)	$(Note 4)	(Note 5) days	(Note 6) days	(Note 7) days
Note 1	$(Note 2)	$(Note 3)	$(Note 4)	(Note 5) days	(Note 6) days	(Note 7) days
Note 1	$(Note 2)	$(Note 3)	$(Note 4)	(Note 5) days	(Note 6) days	(Note 7) days
Note 1	$(Note 2)	$(Note 3)	$(Note 4)	(Note 5) days	(Note 6) days	(Note 7) days
Note 1	$(Note 2)	$(Note 3)	$(Note 4)	(Note 5) days	(Note 6) days	(Note 7) days
Total Project	$(Note 8)	$(Note 9)	$(Note 10)	(Note 11) days	(Note 12) days	(Note 13) days

Note: This table shows the baseline cost, minimum acceptable cost, and maximum acceptable cost for each work package plus the baseline, minimum, and maximum acceptable costs for the total project. It also shows the baseline duration, minimum acceptable duration, maximum acceptable duration for every work package, and the baseline duration, and minimum and maximum acceptable duration times for the total project.

Instructions:

1. List the titles or WBS numbers for every work package.
2. List the budget for each work package. Get this from the budget document. (If work package costs have been estimated, for example, at the 95% confidence level, use these as the baseline costs.) For Notes 3, 4, 6, and 7, see activity 3.
3. Indicate the minimum acceptable cost for each work package. WP minimum acceptable cost = WP cost baseline (same number as Note 2) ÷ upper control limit for cost.
4. Indicate the maximum acceptable cost for each work package. WP maximum acceptable cost = WP cost baseline (same number as Note 2) ÷ lower control limit for cost.
5. List the duration baselines for each work package. This is shown on the work package work order for each work package and on the network diagram. (If work package durations have been estimated, for example, at the 95% level, use these as the baseline durations.)
6. Indicate the minimum acceptable duration time for each work package. WP minimum acceptable duration = WP duration baseline ÷ upper control limit for schedule.
7. Indicate the maximum acceptable duration time for each work package. WP maximum acceptable duration = WP duration baseline ÷ lower control limit.
8. Sum the numbers in the cost baseline column and indicate this sum in the Note 8 space.
9. Sum the numbers in the minimum acceptable cost column and indicate this sum in the Note 9 space.
10. Sum the numbers in the maximum acceptable cost column and indicate this sum in the Note 10 space.
11. Indicate the total project baseline duration in the Note 11 space. The project baseline duration is shown in the network diagram; it is the number above the finish box.
12. Indicate the minimum acceptable duration for the total project in the Note 12 space. (See activity 2.)
13. Indicate the maximum acceptable duration for the total project in the Note 13 space. (See activity 2.)

Maximum acceptable project cost: ------→	$425,000	
Baseline project cost: ------→	$400,000	
Minimum acceptable project cost: ------→	$375,000	

Maximum acceptable project duration: ------→	130 days	
Baseline project duration: ------→	120 days	
Minimum acceptable project duration: ------→	110 days	

Cost upper control limit:	1.066667
Cost lower control limit:	0.941176

Duration upper control limit:	1.090909
Duration lower control limit:	0.923077

A	B Work Package Number	C Baseline Cost	D Minimum Accept. Cost	E Maximum Accept. Cost	F	G Baseline Duration	H Minimum Accept. Duration	I Maximum Accept. Duration
	1	40,000	37,500	42,500		15	13.75	16.25
	2	25,000	23,438	26,563		10	9.17	10.83
	3	60,000	56,250	63,750		14	12.83	15.17
	4	50,000	46,875	53,125		12	11.00	13.00
	5	20,000	18,750	21,250		6	5.50	6.50
	6	30,000	28,125	31,875		8	7.33	8.67
	7	50,000	46,875	53,125		24	22.00	26.00
	8	35,000	32,813	37,188		12	11.00	13.00
	9	40,000	37,500	42,500		10	9.17	10.83
	10	50,000	46,875	53,125		9	8.25	9.75
Total Project		400,000	375,000	425,000		120	110	130

Figure 10.1 Spreadsheet example of a variation control matrix. This template automates activity 3. *Instructions:* (1) In cell F2, enter the maximum acceptable project cost. (2) In cell F3, enter the baseline project cost. (3) In cell F4, enter the minimum acceptable project cost. (4) In cell L2, enter the maximum acceptable project duration. Do not include units of time: days or hours. Show units in cell M2. (5) In cell L3, enter the baseline project duration. Do not include units of time: days or hours. Show units in cell M3. (6) In cell L4, enter the minimum acceptable project duration. Do not include units of time: days or hours. Show units in cell M4. (7) In cells B17 through B26, enter the work package numbers for every work package. Add additional rows as necessary. (8) In cells C17 through C26, enter the work package baseline costs. If the total project budget is based upon work package budgets estimated at 95% confidence, then use these work package budgets. (9) In cells G17 through G26, enter the work package baseline durations. Do not include units of time: days or hours. (10) Note that the total project baseline (cell G30) and minimum and maximum acceptable durations (cells H30 and I30, respectively) are not the sums of their respective columns. These numbers were transferred by Excel from spaces L3, L4, and L2, respectively.

Maximum acceptable project cost: ——→	Note 1		Maximum acceptable project duration: ——→	Note 4
Baseline project cost: ——→	Note 2		Baseline project duration: ——→	Note 5
Minimum acceptable project cost: ——→	Note 3		Minimum acceptable project duration: ——→	Note 6

Cost upper control limit:	#VALUE!		Duration upper control limit:	#VALUE!
Cost lower control limit:	#VALUE!		Duration lower control limit:	#VALUE!

A	B Work Package Number	C Baseline Cost	D Minimum Accept. Cost	E Maximum Accept. Cost	F	G Baseline Duration	H Minimum Accept. Duration	I Maximum Accept. Duration
	Note 7	Note 8	#VALUE!	#VALUE!		Note 9	#VALUE!	#VALUE!
	Note 7	Note 8	#VALUE!	#VALUE!		Note 9	#VALUE!	#VALUE!
	Note 7	Note 8	#VALUE!	#VALUE!		Note 9	#VALUE!	#VALUE!
	Note 7	Note 8	#VALUE!	#VALUE!		Note 9	#VALUE!	#VALUE!
	Note 7	Note 8	#VALUE!	#VALUE!		Note 9	#VALUE!	#VALUE!
	Note 7	Note 8	#VALUE!	#VALUE!		Note 9	#VALUE!	#VALUE!
	Note 7	Note 8	#VALUE!	#VALUE!		Note 9	#VALUE!	#VALUE!
	Note 7	Note 8	#VALUE!	#VALUE!		Note 9	#VALUE!	#VALUE!
	Note 7	Note 8	#VALUE!	#VALUE!		Note 9	#VALUE!	#VALUE!
	Note 7	Note 8	#VALUE!	#VALUE!		Note 9	#VALUE!	#VALUE!
Total Project			#VALUE!	#VALUE!				

Figure 10.2 Spreadsheet template for variation control matrix. This template automates activity 3. *Instructions*: (1) In cell F2, enter the maximum acceptable project cost. (2) In cell F3, enter the baseline project cost. (3) In cell F4, enter the minimum acceptable project cost. (4) In cell L2, enter the maximum acceptable project duration. Do not include units of time: days or hours. (5) In cell M2. (5) In cell L3, enter the baseline project duration. Do not include units of time: days or hours. Show units in M3. (6) In cell L4, enter the minimum acceptable project duration. Do not include units of time: days or hours. Show units in cell M4. (7) In cells B17 through B26, enter the work package numbers for every work package. Add additional rows as necessary. (8) In cells C17 through C26, enter the work package baseline costs. If the total project budget is based upon work package budgets estimated at 95% confidence, then use these work package budgets. (9) In cells G17 through G26, enter the work package baseline durations. Do not include units of time: days or hours. Note that the total project baseline (cell G30) and minimum and maximum acceptable durations (cells H30 and I30, respectively) are not the sums of their respective columns. The numbers were transferred by Excel from spaces L3, L4, and L2, respectively.

Table 10.3   Example of Project Performance Matrix

Work Package (WP)	WP Budget (PV)	Earned Value (EV)	Actual Cost (AC)	Cost Variance (CV)	Cost Performance Index (CPI)	Schedule Variance (SV)	Schedule Performance Index (SPI)
A**	$10,000	$9,000	$9,000	$0	1.000	–$1,000	.900
B	$20,000	$20,000	$22,000	–$2,000	*.909	$0	1.000
C**	$5,000	$4,000	$5,000	–$1,000	*.800	–$1,000	*.800
D	$25,000	$25,000	$26,000	–$1,000	.962	$0	1.000
E**	$30,000	$25,200	$24,000	$1,200	1.050	–$4,800	*.840
Remaining PV	$410,000						
Original project budget	$500,000						
Cum (as of data date)	PV	EV	AC	CV	CPI	SV	SPI
	$90,000	$83,200	$86,000	–$2,800	.967	–$6,800	.924
Critical path cum	PV	EV				SV	SPI
	$45,000	$38,200				–$6,800	*.849

*CPI or SPI out of control; ** Work package on the critical path.

Note: For a discussion of the example, see activities 6 and 7. Figure 10.3 is the automated version.

$1,000, respectively. These cost overruns are out of control — beyond acceptable variance because the CPI indices of .909 and .800 are below the lower cost control limit of .952. Work package C and work package E are $1,000 and $4,800 worth of work behind schedule, respectively. Both incidences of lateness (.800 and .840, respectively) are less than the lower limit of .870. These work packages are late by more than what is acceptable. Work package E is under budget by $1,200 (CV = $1,200). This is acceptable because the CPI of 1.050 is within the acceptable limits of .952 and 1.176.

At day 25, this project is over budget by $2,800 (CV = –$2,800), but this variance is acceptable because the CPI of .967 falls within the CPI limits. The total project is $6,800 worth of work behind schedule (SV = –$6,800). This is also acceptable because the corresponding SPI = .924 falls within acceptable SPI limits: .870 to 1.176.

Figure 10.3 is the Excel© version of Table 10.3. This version can be developed via the template in Figure 10.4.

Activity 7: Determine Critical Path SPI

1. Sum the EVs for work packages on the critical path; this is given the symbol EV_{cp}, reminding us that it is the earned value along the critical path. In Table 10.3, work packages A, C, and E are on the critical path. The EV_{cp} for these work packages is $38,200 ($9,000 + $4,000 + $25,200).
2. Sum the PVs for work packages on the critical path; this is given the symbol PV_{cp}, reminding us that it is the planned value along the critical path. From Table 10.3, total PV_{cp} on the critical path is $45,000 ($10,000 + $5,000 + $30,000).
3. Determine critical path schedule performance index (SPI_{cp}):

$$SPI_{cp} = EV_{cp} \div PV_{cp} = .849 \ (\$38,200 \div \$45,000) \tag{10.10}$$

For Table 10.3, the critical path SPI of .849 is out of control. A schedule variance (SV = –$6,800) on the critical path is beyond what is acceptable; this does not portend well for completing the project within acceptable schedule variance. This project is in control for cost (CPI = .967) and when we consider all the work packages that have been performed, it looks like the schedule is in control (SPI = .924). This is misleading because it is the SPI along the critical path that will determine the project final duration. The SPI_{cp} of .849 does not fall within acceptable limits. Unless the project manager can improve schedule performance along the critical path, this project will be completed with lateness beyond the maximum allowed.

Earned value report as of June 1, 2007 data date

Work Package (WP)	WP Budget (Total PV)	PV as of Data Date	% of WP Complete (as decimal)	Earned Value (EV)	Actual Cost (AC)	Cost Variance (CV)	Cost Performance Index (CPI)	Schedule Variance (SV)	Schedule Performance Index (SPI)
A**	$10,000	$10,000	0.9	$9,000	$9,000	$0	1.000	-$1,000	0.900
B	$20,000	$20,000	1	$20,000	$22,000	-$2,000	0.909	$0	1.000
C**	$5,000	$5,000	0.8	$4,000	$5,000	-$1,000	*0.800	-$1,000	*0.800
D	$25,000	$25,000	1	$25,000	$26,000	-$1,000	0.962	$0	1.000
E**	$30,000	$30,000	0.84	$25,200	$24,000	$1,200	1.050	-$4,800	*0.840

Remaining PV $410,000

OPB $500,000

Cum (as of data date)	PV $90,000	EV $83,200	AC $86,000	CV -$2,800	CPI 0.967	SV -$6,800	SPI 0.924

Critical Path Cum	PV $45,000	EV $38,200				SV -$6,800	SPI *0.849

Latest estimate of project cost (LEPC) = $517,063

Original estimate of project duration (OEPD) = 100 days   Latest estimate of project duration (LEPD) = 118 days

Figure 10.3   Example of spreadsheet project performance report. (See activities 6, 9, and 10.) This figure is the automated version of Table 10.3. *CPI or SPI out of control; (See Activity 2.) **Work packages A, C, and E are on the critical path.

Earned value report as of   Note 10

A Work Package (WP)	B WP Budget (Total PV)	C PV as of Data Date	D % of Work Package Complete	E Earned Value (EV)	F Actual Cost (AC)	G	H Cost Variance (CV)	I Cost Performance Index (CPI)	J Schedule Variance (SV)	K Schedule Performance Index (SPI)	L
Note 1	Note 2	Note 3	Note 4	#VALUE!	Note 5		#VALUE!	#VALUE!	#VALUE!	#VALUE!	
Note 1	Note 2	Note 3	Note 4	#VALUE!	Note 5		#VALUE!	#VALUE!	#VALUE!	#VALUE!	
Note 1	Note 2	Note 3	Note 4	#VALUE!	Note 5		#VALUE!	#VALUE!	#VALUE!	#VALUE!	
Note 1	Note 2	Note 3	Note 4	#VALUE!	Note 5		#VALUE!	#VALUE!	#VALUE!	#VALUE!	
Note 1	Note 2	Note 3	Note 4	#VALUE!	Note 5		#VALUE!	#VALUE!	#VALUE!	#VALUE!	
Note 1	Note 2	Note 3	Note 4	#VALUE!	Note 5		#VALUE!	#VALUE!	#VALUE!	#VALUE!	
Note 1	Note 2	Note 3	Note 4	#VALUE!	Note 5		#VALUE!	#VALUE!	#VALUE!	#VALUE!	
Note 1	Note 2	Note 3	Note 4	#VALUE!	Note 5		#VALUE!	#VALUE!	#VALUE!	#VALUE!	
Note 1	Note 2	Note 3	Note 4	#VALUE!	Note 5		#VALUE!	#VALUE!	#VALUE!	#VALUE!	
Note 1	Note 2	Note 3	Note 4	#VALUE!	Note 5		#VALUE!	#VALUE!	#VALUE!	#VALUE!	
Note 1	Note 2	Note 3	Note 4	#VALUE!	Note 5		#VALUE!	#VALUE!	#VALUE!	#VALUE!	

Remaining PV   Note 6

OPB   #VALUE!

Cum (as of data date)   PV $0   EV #VALUE!   AC $0   CV #VALUE!   CPI #VALUE!   SV #VALUE!   SPI #VALUE!

Critical Path Cum   PV Note 7   EV Note 8   Note 9   SV #VALUE!   SPI #VALUE!

Latest estimate payout cost (LEPC) =   #VALUE!

Original estimate of project duration (OEPD) =   Note 9

Latest estimate of project duration (LEPD) =   #VALUE!

Figure 10.4 Spreadsheet template for project performance report. Notes on next page.

Figure 10.4   Spreadsheet template for project performance report. (See activities 6, 7, and 8.)

Instructions:

(1) In the Note 1 space (A7 through A16), enter the work package number for all work packages that have been started.

(2) In the Note 2 space (B7 through B16), enter the total budget for every work package listed in Note 1.

(3) In the Note 3 space (C7 through C16), enter the dollar value of work scheduled to be completed by the data date. The total PV equals the PV as of the data date unless the work package is not scheduled to be completed by the data date. In this case, PV as of data date equals the dollar value of the work scheduled to be completed by the data date.

(4) In the Note 4 space (D7 through D16), enter the percent of the total PV completed by the data date. Enter the number as a decimal: .75, not 75%.

(5) In the Note 5 space (F7 through F16), enter the actual cost of each work package as of the data date.

(6) In the Note 6 space (B19), enter the amount of work (in dollars) remaining to be completed after the data date.

(7) Enter a double asterisk (**) to each work package number that is on the critical path. Sum the PV as of the data date for all of these work packages. Enter this sum in the Note 7 space (C28).

(8) Sum all the earned values in column E for work packages that are on the critical path. Enter this sum in the Note 8 space (E28).

(9) In the Note 9 space (E32), enter the original estimate of project duration. This is the number over the final box on the network diagram.

(10) Enter the current date in the Note 10 space (L1).

Activity 8: Forecast Project Final Duration

Project final duration [called latest estimate of project duration (LEPD)] equals the original estimate of project duration (OEPD) divided by the critical path schedule performance index (SPI_{cp}):

$$LEPD = OEPD \div SPI_{cp} \tag{10.11}$$

The forecasted duration, that is, the latest estimate of project duration, is 118 days (100 days ÷ .849). Please note that forecasting final project duration requires that you determine the SPI for the critical path because it is the critical path that determines the project duration. The maximum acceptable project duration is 115 days, but the SPI_{cp} forecasts the completion of the project three days beyond the maximum acceptable duration.

Activity 9: Forecast Project Final Cost[6]

The cumulative CPI can be used to forecast the project's final cost. The latest estimate of project cost (LEPC) is a forecast of project final cost. LEPC equals the original estimate of project cost (OEPC) divided by the project's cumulative CPI:

$$LEPC = OEPC \div CPI \tag{10.12}$$

Thus we can forecast a final project cost of $517,063 ($500,000 ÷ .967). This final cost will be $7,937 below the maximum acceptable cost of $525,000.

How to Control the Project

Up to this point, this chapter has discussed ways to establish metrics with which to measure how well the project is being controlled. Measuring how well a project is being controlled is not the same thing as performing the actions that keep the project "in control." Control means keeping variances within acceptable limits. The actions taken by the project manager to do this start in the initiation phase and continue through the planning and execution phases. The remainder of the chapter will discuss these actions.

Control during the Initiation Phase

Controlling variance starts in the initiation phase with the imposition of discipline in determining the problem or requirements driving the need for the project. Ill-defined or incomplete documentation of requirements means that the nature of the deliverables will be in flux and subject to varying interpretation. Achieving the efficiencies implied by keeping variances small requires a clear and unchanging target. Unclear and indeterminate requirements subject the project to numerous scope changes and increased risk. Scope changes produce schedule and cost variances even when the scope changes are properly managed. Dealing with risk events costs time and money, thereby producing the kinds of variances we seek to limit.

A second major activity of the initiation phase is defining the project. The purposes of the business case definition process are threefold:

1. To clarify the project for the project team
2. To provide management with the information with which to make an informed decision about whether to invest in the project
3. To assure that the project team and the sponsor have the same understanding of the project

Recovering from a lack of clarity will cost money and time.

> **Example 8:** The project manager assumes that the customer will provide physical security for the building materials, while the customer assumes that the project manager will provide this security. The theft of these materials will produce cost and schedule variances. At a minimum, the business case definition should include the following topics: needs or problems that the project is to solve, major project phases and major activities, deliverables, assumptions, constraints, out of the ordinary resource requirements (i.e., a specialized skill set, material or equipment that the project team will have to make a special effort to obtain), potential problems, project budget, duration targets, and any customer-imposed constraints or special requirements. A lack of clarity and agreement on these topics will produce unwanted variances. The project should not move into the planning phase if the business case definition has not been completed and approved by the sponsor. Defining the project and getting the definition approved should be the first team task if this has not been previously completed.

Control during the Planning Phase

Although project control starts in the initiation phase, the actions taken by the project manager during the planning phase predispose the project to successful or unsuccessful control. It is what the project manager does during the planning phase that will make control achievable. Project manager behaviors that affect project controllability fall into the following categories:

1. Manage risk
2. Discipline time and cost estimates
3. Clarify scope
4. Include risk in the schedule and budget estimates
5. Plan and coordinate resources

Manage Risk

Dealing with risk events costs time and money. These costs and lost time produce unwanted variance. A rigorous risk management process will limit the effect of unidentified or uncontrolled threats to the project. Risk events identified early are usually easier to avoid or reduce in impact; threats identified early are usually less costly to deal with than those identified later. (For information on how to develop the risk management plan, see Chapter 9.)

Discipline Time and Cost Estimates

Viable estimates are achieved when the following conditions are met:

1. Estimates come from conscientious people who have direct experience in the work they are estimating.
2. Estimators are held accountable for the accuracy of their estimates.
3. Estimators have a complete understanding of the scope of the work package and the circumstances and conditions under which the work will be performed.
4. Acceptable estimating methods are employed.
5. The assumptions associated with the estimates are explicitly stated and they are acceptable.

One way to encourage the use of an acceptable estimating method is to establish a procedure and format for soliciting estimates. Table 5.3 is an example of a work package estimate sheet. (Templates are on Table 5.1 and Table 5.4.) Its purpose is to facilitate viable estimates. It clarifies work package scope, assumptions, constraints, and risks. The worksheet encourages the project manager and estimator to be complete in providing this information. The project manager initiates the form and describes the scope of the work package including subtasks and deliverables. The estimator lists assumptions, constraints, and risks associated with the work package. The estimator lists the subtasks and deliverables.

If desired, the PERT effort time estimates for completing each subtask or deliverable and the availability of the person who will perform the work package are included. Availability (e.g., full time versus half time, etc.) is a condition that determines duration time. The estimator lists all materials needed with their cost as well as any equipment required including its cost if the project manager is to pay for these. On construction projects, subcontractors often provide a total labor and materials firm-fixed price; this is common for tile, electrical, and plumbing contractors in the home-building industry.

To PERT or Not to PERT

When using the work package estimate sheet, the project manager has the choice of requiring or not requiring the use of the PERT method of estimating effort time. Advantages of the PERT method are:

1. More accurate estimate of effort time
2. The ability to estimate the standard deviation
3. The ability to add a reserve to deal with risk into estimates of effort time and labor cost

(For a discussion of the PERT method, see activity 1 in Chapter 5.)

The project manager has three options in estimating effort times and labor costs:

1. Based on a single estimate of effort time (see Table 5.1)
2. Based on the average effort time (see Table 5.3)
3. Based on the effort times at the 90% or 95% levels of confidence (see Figure 5.1)

One of the biggest reasons for cost and schedule overruns is naïve estimates of work package effort time and cost. Using work package effort times and labor costs estimated at the 90% and 95% levels will certainly reduce schedule and cost variances, thereby directly helping to control the project.

The estimator and the estimator's supervisor sign and date the estimate worksheet. Requiring the estimator and supervisor to sign the estimating sheets is important. It sends a strong signal that estimators are responsible for their estimates. And this is important. Viable estimates are essential to project control, and getting viable estimates depends upon getting conscientious people to perform the estimates. Table 5.5 is an example of the bottom-up technique; Table 5.6 is the template. This is the technique of choice when determining the project's budget and schedule baselines.

Unacceptable Alternative Methods

The parametric and analogous methods described below are not acceptable methods for determining the project's budget baseline. The parametric and analogous strategies produce quick estimates that are subject to large errors because both are based on generalities with unspecified assumptions. These techniques do not include the specifics of the work package scope. They are acceptable during the initiation phase when a "ballpark" approximation is desired but not acceptable when a budget baseline is needed.

> **Example 9. The parametric method:** Last year our firm constructed a 30 foot × 40 foot basement room. The room included 2 × 4 construction, half-inch sheetrock walls, and ceiling. The floor was constructed of three-quarter-inch oak flooring. The room included one entry door plus a 3 foot × 8 foot closet outside the 30 × 40 footprint. It included electrical receptacles every 8 feet plus a single switch to control the three two-light, 4-foot overhead fluorescent light fixtures. Total labor and material cost was $13,071.

Our task now is to estimate the cost of a 10 feet × 20 feet room to be constructed in the basement of a new house. The parametric method would employ a single equation to estimate the cost of this room. With the information about the 30 feet × 40 feet room we constructed last year, we may develop a single equation (parametric) with which to estimate the total cost of this smaller room. From the information in example 9, we know that it cost $13,071 to construct a room with 1,200 square feet. Thus it cost $10.89 per square foot (cost = $13,071/1,200 square feet = $10.89 per square foot). The parametric method would use this equation to estimate the cost of the 200 square foot room as $2,178 (10.89 per square foot × 200 square feet). The parametric method ignores the significant differences in the specifics of these two rooms. The differences in the materials and the methods of installing these different materials will result in significant differences between the estimated cost and the actual cost. Because the parametric method provides a rough approximation of the cost, it should not be used to establish the cost baseline (project budget). To avoid the error inherent in using one equation to estimate the cost and the resulting unwanted variance (error), the project manager should not employ parametric estimates or accept estimates that employ this technique.

> **Example 10. The analogous method:** The analogous method is another technique that produces a rough approximation of cost. This technique uses the costs for large elements of cost of a previous project with which to estimate the cost of a similar project. The analogous estimate is shown in the following table:

Previous Project		Work Package Being Estimated	
Baseline cost	$13,071	Baseline cost	$13,071
		Differentials	
Materials	$9,071	−5/6($9,071)[a] =	−7,559
Labor	$4,000	−5/6($4,000)[a] =	−3,333
Analogous estimate of 200 square foot room = $2,179			

[a] Materials and labor costs are assumed to be five-sixths less than those of the previous project because a 200 square foot room is one-sixth the size of the 1,200 square foot room (200/1,200 = 1/6). Hence these costs are assumed to be five-sixths less than those of the larger room. (Note that the analogous method provided an estimate similar to the parametric method.) This work package will produce an actual cost that is significantly different from the estimate because it does not break the costs into small enough elements of work and materials. The analogous method is also not accurate enough for determining the work package cost baseline.

Clarify Scope

Viable estimates of effort time and cost depend upon a clear understanding of the total scope for each work package. An incomplete understanding of the subtasks that must be performed will obviously result in a naïve (too low) estimate of effort time and cost. The work package scope portion of the estimate sheet is used to assure a complete understanding of the work package scope. (See Table 10.4.)

Build a Schedule and Budget That Include a Reserve for Risk

The PERT method includes equations that provide an easy way to build a reserve into the estimates of effort time and work package cost. The average effort time (t_e) to perform a work package is Equation 5.1:

$$t_e = [P + 4(ML) + Op] \div 6$$

where P is the pessimistic or longest estimate, ML is the most likely estimate, and Op is the optimistic or shortest estimate. The standard deviation for a work package is Equation 5.2:

$$SD = (P - Op) \div 6$$

An easy and conservative way to include a reserve for estimates of effort time is to determine the maximum effort time at 95% confidence level, as given in Equation 5.3:

$$\text{Max time}_{.95} = Te + 2\sigma$$

Table 10.4   Example of a Work Package Estimating Worksheet

Project Name: Basement Room   **Work Package #/Name:** 2.1 Install Wallboard

Project Manager: Lou Passo   **Date:** Dec 12, 2005

Work Package Scope: Install half inch wallboard (sheetrock) on walls and ceiling in a 10' × 20' room with 8' ceiling. Spackle all joints and fastener holes and sand all surfaces so walls and ceiling are ready for painting. Install plastic corner beads on all external corners. Install joint tape on all joints including internal corner joints. Room entrance is an open arch without a door from the descending stairwell. Room will have previously been studded out with 2" × 4" wood stud construction. Rough wiring will have been previously installed for eight wall receptacles and two overhead light fixtures.

Assumptions/Constraints/Risks: All material including wallboard, corner bead, joint tape, joint compound, fastener screws, and sandpaper will be located in the basement adjacent to the new room. Workers provide their own tools.

Subtasks/Deliverables	# People	LLR	PERT Effort Time			Avg Time (hours)	Std Dev	Avg Cost	.95 Time (hours)	.95 Cost (hours)
			Op	ML	Pess					
Install wallboard	2	$40t	22	24	30	24.7	1.3	$988	27.3	$1,092
Apply joint tape & joint compound	2	$40t	5	8	12	8.2	1.2	$328	10.6	$424
Prepare walls & ceiling for painting	2	$40t	8	12	16	12	1.3	$480	14.6	$584
Total Average Time and Average Cost						44.9		$1,796		
Total Time and Cost At 95% Confidence									52.5	$2,100

Materials List

Description	Quantity	Unit Cost	Item Cost
1/2 inch sheetrock (4' × 8' sheets)	33	$6.50	$214.50
External corner bead (8' long)	3	$5.00	$15.00
Joint compound (5 gallon can)	1	$25.00	$25.00
#8 × 1-1/4 wallboard screws	5 lbs	$9.00	$9.00
Joint tape rolls	5	$2.00	$10.00
Sandpaper sheets	10	$.65	$6.50
Total Materials Cost			$280.00

(Continued on next page)

Unusual Equipment Required

Description	Quantity	Unit Cost	Item Cost
None			

Total Equipment Cost $0

Agreement: These estimates of effort times and labor costs are based on the work being performed by Jim Smith and Manuel Jones or someone else who can complete the work in the same estimated time. The work package will be completed on a full-time basis unless stated otherwise in the comments below.

Estimator: Jim Smith   **Date:** Dec 15, 2005
Supervisor: Mary Cooper   **Date:** Dec 16, 2005
Comments:

Project Manager Calculations — Summary of Estimates

	Average	$Max_{.95}$		
Labor Cost	$1,796	$2,100	Total t_e	44.9 hours
Material Cost	$280	$280	Availability	100%
Equipment Cost	$0	$0	Duration Time	44.9 hours
Total Cost (Avg/Max .95)	$2,076	$2,380	Effort Time$_{.95}$	52.5 hours
			Duration Time$_{.95}$	52.5 hours

PERT Equations

PERT average effort time $(t_e) = (P + 4[ML] + Op) \div 6$

Standard deviation $= (P - Op) \div 6$

Duration time $= t_e \div$ Availability

Note: t = Loaded labor rate (LLR) is for both workers combined.

Table 10.4 continued.

Table 10.4 shows the average effort time as 45 hours; The max effort time at 95% level of confidence = 52.5 hours. The project manager has the option of using the average effort time of 45 hours or the maximum time of 52.5 hours in the network diagram as the duration for the work package. Forty-five hours does not include a risk reserve in the estimate whereas 52.5 hours does include a reserve. You can be 50% confident that the work package will be completed in 45 hours, or 95% confident that it can be completed in 52.5 hours. What is the more prudent strategy: using estimates in which you are 50% confident or estimates in which you are 95% confident? Of course, using the estimates at 95% confidence will produce longer project duration than one based on estimates at the 50% level of confidence. If you use an effort time estimate of 45 hours and the work package is completed in 52.5 hours, you have experienced 7.5 hours of negative variance, i.e., the work package was completed 7.5 hours late. If you use an effort time estimate of 52.5 hours and the work package is completed in 45 hours, you have experienced 7.5 hours of positive variance, i.e., the work package has been completed 7.5 hours early! Please note that the purpose of all project control is to keep all variances, including schedule variance, as small as possible or positive (i.e., coming in early or under budget) if possible. Project schedules based on the average time to complete work packages underestimate the actual time it will take to complete the project; these projects will typically come in late. Projects in which work package durations are based on effort time estimates at the 95% level of confidence provide a conservative (longer duration time) estimate of how long the project will take. These projects have a high probability of completing on time or early. The template in Figure 5.1 gives the project manager the ability to estimate effort times and labor costs at the 50%, 90%, and 95% levels of confidence. Estimates at the 90% and 95% levels provide estimates that are easier to meet than those based on 50% confidence.

Work package cost consists of three components: labor, material, and equipment. The risks of material and equipment increases are based upon estimates of inflation. For relatively short-term projects, one year or less, inflation will usually have a minor impact. Labor costs are easy to determine because this cost is set by effort time. Table 10.4 indicates a loaded labor rate (LLR) for the two-person crew of $40 per hour:

The average labor cost = average effort hours × LLR.
The average labor cost = 44.9 hours × $40/hour or $1,796.
Total work package cost = labor cost + material cost + equipment cost.
The average total cost = $1,796 + $280 + 0 = $2,076.
The labor cost at 95% confidence level = 52.5 hours × $40/hour = $2,100.
The total work package cost at 95% confidence level = $2,100 + $280 + 0 = $2,380.

Again, the project manager has the choice of using the lower estimate of $2,078 (in which the project manager has 50% confidence) or using $2,380 for which the project manager is 95% confident. The lower estimate does not include the cost associated with the risk of the work package taking longer than estimated to complete, whereas the higher estimate includes the risk of the work package taking longer to complete. (The template in Figure 5.1 and Table 10.4 give the project manager the ability to estimate labor costs at 50%, 90%, and 95% levels of confidence.)

Develop a Resource Plan

During the execution phase, the project manager's primary function is to coordinate the efforts of others. This coordination can be stated as:

> Assuring the right person shows up at the right place, at the right time, with the right understanding of what has to be accomplished, with the right material, equipment, and budget to accomplish the work package.

Any break in this chain will produce unwanted lateness and increased cost. If the worker does not show up at the right time, the project suffers a schedule delay (i.e., schedule variance). If the worker shows up on time but the material is not available, there is a delay until the material arrives. This will produce both schedule and cost variance. If the worker shows up on time and the material is at hand but the worker does not have a correct or complete understanding of work package scope, the project will experience schedule and cost variances produced by the effort it takes to correct the misunderstanding and fix the error. If the worker shows up on time, has the material needed, has a complete understanding of the work package scope but the equipment is not available, the project will suffer schedule and cost variances until the equipment arrives.

The resource plan is the single document that can help the project manager keep track of all this information and coordinate these activities aggressively and proactively. The resource plan summarizes the information that exists on the other planning documents. It includes the names of all work packages; who will do each work package (i.e., the individual, department, or subcontractor); the early start and early end dates for each work package; and the material, equipment, and budget for each work package. It also includes comments that remind the project manager of actions that must be taken for each work package. Table 10.5 is an example of a resource plan for a single work package. Table 7.1 is a format for the resource plan.

Table 10.5   Partial Example of a Resource Plan

Work Package:	2.1 Install wallboard
Who:	J. Smith and M. Jones
Start to end dates:	9/4/07 to 9/11/07 (52.5 hours) (Sept 8 and 9 are non-work days)
Budget:	$2100 labor + $280 material = $2380
Materials & supplies	
	33 sheets ½ inch wallboard (4′ × 8′)
	3 exterior corner beads
	5 gal joint compound
	5 lbs fasteners
	5 rolls joint tape
	10 sheets sandpaper
Equipment:	Low height scaffolding
Comments:	Room temperature must be above 45 degrees Fahrenheit.

Note: Example shows the information used by the project manager to coordinate all the resources required for a single work package.

Control during the Execution Phase

The project manager's actions during the execution phase are aimed at assuring that the project is performed in accordance with the plan. The resource plan will be of considerable help in aggressively and proactively coordinating the efforts of others to assure that the project is executed as planned. This means lots of telephone calls or email messages placed a week or two before each work package is scheduled to assure that the right person shows up at the right place, etc.

A second primary concern for the project manager is the effects of threat events that produce unwanted schedule variance (lateness) and cost variance (additional costs). Continuous monitoring of identified threats and continuous attempts to identify new threats is how the project team keeps the effects of threats to a minimum. The efforts to recover from threat events cost time and money. Strategies that avoid, mitigate, or transfer risks are strategies that reduce the cost and schedule variances caused by the threat events. (Dealing with project risk is covered in Chapter 9.)

A third concern during the execution phase is to control and document scope changes. Determining and documenting the cost and schedule impact of scope changes are an important part of control. Any scope change for which the project is not reimbursed produces unwanted schedule and cost variance. Even scope changes that are reimbursed cause schedule and cost variance. They disrupt the project's ongoing work and require the team to reorganize its activities and resources. (Scope management is discussed in Chapter 11.)

This chapter asserts that the purpose of all control is to keep scope, cost, and schedule variances within acceptable limits or as small as possible. It listed the behaviors required of the project manager and project team if their efforts at minimizing variance are to be successful. The discipline and attitude necessary to control variance may not be typical of the organization's culture. In this instance, the project manager faces additional difficulty in controlling variance because without the discipline outlined in this chapter and the attitude implied by the discipline, project control is nearly possible.

Notes

1. The content of Chapter 10 is adapted from the following articles written by the author: "How to Set and Use Project Control Limits" (*ESI Horizons,* March 2006) and "How to Set and Use Project Control Limits, Part II: How to Control Your Project" (*ESI Horizons,* June 2006). © ESI International, 901 North Glebe Road, Suite 200, Arlington, VA 22203; www.esi-intl.com. With permission.)

2. The upper and lower thresholds of 76 and 70 degrees Fahrenheit are used for illustration only. The actual thresholds differ depending upon the temperature-sensing technology. Bimetal thermostats are less sensitive than are electronic thermostats, so the upper and lower thresholds of the former would be further from the desired temperature than those of the electronic thermostats.

3. Setting a desired lower limit on cost seems like an oxymoron but there are times when this applies; e.g., on a cost plus contract where a minimum cost by the contractor is necessary to maintain profit margin. Also work packages with very low costs are a concern because the low cost may be the result of incomplete performance, low quality, or some other condition that is not desired.

4. An alternate method of establishing cost control limits is to select arbitrarily the percentage of acceptable variation around the cost performance index. For example, we may arbitrarily set the cost performance index at plus or minus 10%. This means that the upper control limit is 1.100 (1.000 + .100) and the lower control limit is .900 (1,000 − .100). If the original budget is $200,000 and we use the limits of .90 to 1.10, the maximum and minimum acceptable costs would be $222,222 ($200,000 ÷ .900) and $181,818 ($200,000 ÷ 1.100), respectively. Thus setting the cost index at plus and minus 10% determines that the maximum acceptable cost will be $222,222 and the minimum acceptable cost will be $181,818. If these maximum and minimum costs are not acceptable, adjust the plus and minus tolerances to establish limits of cost that are acceptable.

5. An alternate method of establishing schedule control limits is to select arbitrarily the percentage of acceptable variation around the schedule performance index. For example, we may arbitrarily set the schedule performance index as plus 5% or

minus 15%. This means that the upper SPI control limit is 1.05 (1.00 + .05) and the lower control limit is .85 (1.00 – .15). Using these limits means that the project maximum duration is set at 118 days (100 days ÷ .85) and the minimum duration is set at 95 days (100 days ÷ 1.05). Again, the plus and minus tolerances may be adjusted to meet the sponsor's requirements.

6. An alternative expression of this equation is: EAC = BAC ÷ CPI, where EAC and BAC refer to estimate at completion (forecasted cost) and budget at completion (meaning the original budget), respectively. (EAC equation is used with permission.)

Chapter 11

Scope Change Procedure

Mega Recipe for Dealing with Scope Change

Activity 1: Establish the scope change procedure.
Activity 2: Communicate the scope change procedure to the stakeholders.
Activity 3: Assign one team member to manage the scope change procedure.

Purposes of the Scope Change Procedure

The purposes of the scope change procedure are to control, document, and communicate changes to the project scope. The project manager is responsible for creating the scope change procedure and seeing that all sponsors, customers, team members, and others use it properly. This chapter outlines a generic change control procedure that can easily be modified to meet particular project needs.

Scope changes may come from anywhere: from sponsors, customers, stakeholders, team members, via government regulations, and changes in environment conditions depending upon the nature of the project. They may arrive as a formal request, by telephone, email, or over a casual cup of coffee. Regardless of the source or manner in which the scope change arrives, it is important that each be processed in a manner consistent with the established procedure.

Role of the Project Manager

The project manager's role in scope change is to assure that a disciplined procedure is used to control, document, and communicate all changes in the project scope. As with other responsibilities, the project manager directs the team in its activities to establish and use the procedure.

It is important that the project manager inform all interested parties of the intent and content of the procedure to preclude any misunderstandings or ill feelings. Early in the initiation phase, the project manager must brief the procedure to all stakeholders, customers, and sponsors to prevent future problems.

Outputs of the Procedure

1. Increased control over what project scope gets changed.
2. Comprehensive documentation concerning the nature, source, and disposition of changes will be produced and information will get to the people who need to be kept informed.
3. The procedure will produce an audit trail of all changes to the project.

Problems Implementing the Scope Change Procedure

The problems that may arise relative to the scope change procedure are related to not having a formalized procedure, not following the procedure, or having too many changes in rapid succession.

The problems associated with not having a procedure or choosing not to use it are really the result of project manager inexperience. However, having and using a disciplined procedure can also cause problems.

Dealing with scope changes produces stress for the project team. Every time a change is initiated, the team must activate the scope change procedure; that is, it must stop what it is doing and resolve the proposed change. Team members have to be reassigned to deal with the scope change. The momentum of performance is disrupted. If a change is approved, all the planning documents and the baseline documents have to be updated. It is easy to fail to update a document that should be changed. For this reason, too many scope changes hitting the project in a short time will quickly overwhelm the team with impacts to determine and documents to change. Imagine the confusion if the team has to deal with numerous changes during the same week or month where the impact and documentation from the previous changes may or may not have been completely processed. In this case, it may be necessary to designate one team member to coordinate scope changes on a full-time basis.

Activity 1: Establish the Scope Change Procedure

Step 1: Document the source, date, and nature of scope change on the scope change control form. Include the name and signature of the requesting official and the date. Figure 11.1 depicts a scope change form with instructions for its use.

Scope Change Request Form

Step 1. Requesting Official state nature of scope change required:

Requesting Official Name: _____ Signature _____ Date _____

Step 2. Project Manager validate that requesting official is authorized to make scope changes.

Validating Official Name: _____ Signature _____ Date _____

Step 3. Estimate of cost and time to accomplish scope change:

Estimate by: _____ Name _____ Signature _____ Date _____

Step 4. Describe the impact of scope change on the schedule end date:

Determination by: Name _____ Signature _____ Date _____

Step 5. Requesting official indicates acceptance or rejection: (Check One)

—Accept additional cost and schedule impacts of this change.
—Reject additional cost and schedule impacts of this change.

Authorizing Official Name: _____ Signature _____ Date _____

Step 6. Project Manager change planning and baseline documents to reflect scope change.

Step 7. Project Manager communicate scope change to appropriate stakeholders.

Step 8. Project Manager annotate change in the Scope Change Log.

Figure 11.1   Scope change request form with embedded procedure.

Step 2: Validate the proposed change by determining that:
1. The request comes from someone authorized to suggest changes, that is someone authorized to pay for the change and authorized to move the project completion deadline. Requests for scope changes coming from an unauthorized person must be forwarded for resolution to the "customer" official authorized to make scope changes.
2. The request is feasible. A change to engineer specifications or software features would be sent to the engineering or software departments for this decision.

Step 3: Determine impact of scope change upon cost and schedule. The following questions must be answered: How much will it cost to accomplish the scope change? How long will it take to accomplish? These questions are answered by the head of the department affected by the change, e.g., the engineering department head for changes to engineering specifications or the chief of software design for any changes to software features. Another question: What is the impact of the change upon the project final delivery date? This question is answered in step 4.

On large and expensive projects, a committee consisting of representatives from the various disciplines within the organization involved in the project would perform step 3. The committee called the change control board (CCB) may or may not include the project manager as a voting or nonvoting member.

Step 4: The project team determines the impact upon schedule by adding the scope change to the precedent diagram and repeating the forward and backward passes. Figure 11.2 and Figure 11.3 show the impact of a scope change that increases the time to complete work package E from seven days to

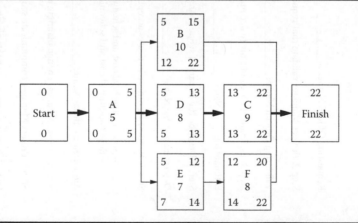

Figure 11.2   Original precedent diagram showing schedule BEFORE scope change. Work package duration times are shown in the center of the rectangle; i.e., duration time for work package A is five days.

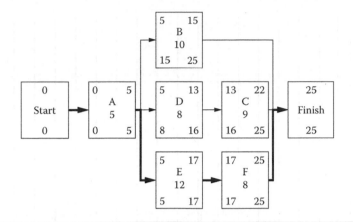

Figure 11.3 Precedent diagram showing schedule AFTER the scope change.
The duration time for work package E is changed from 7 days to 12 days. Schedule impact: project duration time is extended by three days from 22 to 25 and the critical path has moved. Prior to the scope change, the critical path consisted of start, A, D, C, and finish. After the scope change is incorporated, the critical path consists of start, A, E, F, and finish.

12 days. The impact on cost and schedule are then documented on the change request form and signed by the appropriate official.

Step 5: Get the customer's approval of additional cost and revised schedule. This requires the customer's signature and date of signature on the change request form. The approval of the additional time and cost must be signed by someone in the customer organization with the authority to do so. The following is a short, true story that demonstrates the importance of this requirement.

In the late 1990s the U.S. Navy contracted for the construction of a ship, which was to be delivered to the Atlantic fleet in Norfolk, Virginia. After the ship's completion but before delivery, the general contractor conducted a personal tour for the Admiral with acquisition authority at the Norfolk Naval Base. Sometime during the tour, the Admiral ran his hand over the surface of a bulkhead and said, "This is an unusual shade of gray." He said nothing further. The tour ended and the two men parted. Afterward, the general contractor thought that it was advisable to repaint the ship. He had this done at a cost of approximately one million dollars and he submitted a bill to the Navy. The Navy contracting officer denied payment because the change was not authorized. The contractor was never reimbursed for this cost.

Step 6: Change the project planning documents and baseline documents as necessary. For example, if an approved scope change adds a work package to the work breakdown structure (WBS), it will be necessary to change the

Scope Change Log

Request Date	Requesting Official	Description of Change	Cost Impact	Schedule Impact	Final Resolution

Figure 11.4 Scope change log form.

WBS and change the budget, network diagram, Gantt chart, and other planning documents accordingly.

Step 7: Communicate the scope change and its impact upon cost and time to whomever has to be kept informed: stakeholders, sponsor, team members, or others who need to know of the change or will be involved in implementing the change.

Step 8: Annotate the change control log and retain all documentation in a change control folder. The purpose of the change log is to create a summary of all the proposed scope changes and their eventual resolution. This document is useful in seeing at a glance the total impact of all the changes proposed and implemented. The purpose of the change control folder is to retain the detailed information about each scope change, including the scope change form. This information may be needed to resolve discrepancies. Figure 11.4 is an example of the scope change log with instructions on its use.

Activity 2: Communicate the Scope Change Procedure to the Stakeholders

The project manager needs to communicate the date upon which the scope change procedure goes into effect. Usually this occurs immediately after the project baseline is established. The baseline is the customer's expectations of scope, cost, and

---

Memo to:	Ms A. Jones	Date:
	Mr. B. Scrifft	
	Mr. J. Lukor	
From:	John Albertson,	
	Opus 12 Project Manager	
Subject:	Scope Chance Procedure	

It's important that all proposed scope changes are properly controlled, documented, and communicated. To serve these three purposes, we will use the following procedure:

1. Scope change requester submits change request on the form attached (Figure 11.1).
2. The project manager or team member will verify that the scope change comes from someone authorized to make the change.
3. Project manager (or team member) and appropriate department head will determine the impact of the proposed change on cost and schedule. This will be noted on the change request form.
4. The scope change form will be submitted to the customer or sponsor with the request to approve or disapprove payment of the additional cost and schedule time required by the change.
5. With approval, the project manager will have the planning and baseline documents revised.
6. Disposition of the change request will be added to the Opus 12 scope change log. (Figure 11.4)
7. The project manager will see that all necessary stakeholders are informed of the scope change.
8. The project manager will see that the change is implemented and that all documents are filed.

Figure 11.5 Memo announcing the scope change procedure.

time for the project. These are established in the planning phase when the planning documents are developed. Figure 11.5 is an example of a memo announcing the scope change procedure.

Activity 3: Assign Someone to Manage the Scope Change Procedure

The project manager needs to assign one or two people the additional duty to facilitate the scope change procedure.

Chapter 12

Project Closeout Plan

Mega Recipe for Closeout Planning

Activity 1: Perform project closeout.
Activity 2: Perform client closeout.
Activity 3: Perform organizational closeout.
Activity 4: Conduct subcontractor closeout.
Activity 5: Perform final risk assessment.
Activity 6: Write the project final report.
Activity 7: Conduct team closeout.

Purpose of the Closeout Planning Activities

The purpose of the project closeout activities is to end the project in a professional manner — one that reflects well upon the team, the organization, and especially upon the project manager. The purpose of the closeout plan is to assure that the activities will be conducted in an appropriate manner and that the resources (people, money, and time) for the activities will also be available at the end of the project.

The closeout activities are conducted during the project closeout phase, but the planning for these activities should be performed along with the planning for the other project activities. The work packages to be performed during the execution phase were listed in the work breakdown structure (WBS) during the planning phase. The activities to be performed to close the project are legitimate work packages that must also be listed in the WBS. The closeout work packages need to be included in the budget, in the schedule, and in the resource plan. The only difference between the usual work packages and the closeout work packages is that the latter are usually short in duration. Because they are small in duration, one day or less, they have been grouped into seven work packages with subtasks.

Activity 1: Perform Project Closeout

The purpose of this work package is to determine that the project is complete or if there are any requirements that were not completed. The subtasks include:

1. **Validate project completion:** Examine the WBS to determine that all the work packages have been completed. Annotate the WBS to indicate any work packages that are not complete or will not be completed, and indicate the reason. Retain this "final WBS" in the archives.
2. **Review the business case definition and charter:** Indicate any deviations from the desired deliverables listed in these documents. Retain the annotated document for the archives.
3. **Review the contract:** If there is a contract, assure that all the requirements of the contract have been completed. Annotate the contract as necessary to indicate any deviations of performance from that specified in the contract. Retain the annotated contract for the archives.
4. **Conduct a final project evaluation:** Figure 12.1 provides an outline. The information generated by Figure 12.1 will be used in the closeout phase when the final report is written. Figure 12.2 is the outline for the report.

Estimate the time and budget needed to do this work package and include it in the WBS, budget, schedule, and resource plan, etc.

Activity 2: Perform Client Closeout

Two subtasks make up this work package:

1. **Validate deliverable(s) submission and document that all deliverables were accepted by the customer:** If no documentation exists, write a memo to the customer with a retained copy indicating that the deliverables were submitted and the date of the submission. The customer's failure to respond with a notice of rejection is tantamount to acceptance.
2. **Assess customer satisfaction:** It is important to assess the level of customer satisfaction because it is easier to get repeat business from a customer than it is to find another customer. Once you have worked for an organization, you want to leave the relationship with the customer predisposed to use you again. The feedback from the survey will also serve to generate lessons learned. Figure 12.3 is an example of a customer satisfaction survey.

1. Are all work packages complete? If not, explain.
2. Are all requirements of the business case definition and charter met? If not, explain.
3. Have we met all of our customer's contractual requirements? If not, explain.
4. Have our subcontractors met all of our contractual requirements? If not, explain.
5. Have we produced all required deliverables? If not, explain.
6. Have all deliverables been accepted by our customer? If not, explain.
7. Supply the following final project data:
   (a) Original estimate of project cost (OEPC): $_____
   (b) Total budget for approved scope changes: $_____
   (c) Final project total budget: $_____
      (c) = (a) + (b)
   (d) Actual project final cost: $_____
   (e) Cost variance at completion (VAC): $_____
      (e) = (c) – (d)
   (f) Cost of threat recovery: $_____
   (g) Cost of opportunity recovery: $_____
   (h) Total cost of risk recovery: $_____
      (h) = (f) + (g)
   (i) Used risk reserve as percentage of final project total cost: $_____
      (i) = (h) ÷ (d)
8. Project final earned value status:

CV: _____	Max. Acceptable Cost:	$_____	
SV: _____	Min. Acceptable Cost:	$_____	
CPI: _____	Max. Acceptable Duration:	$_____	
SPI: _____	Min. Acceptable Duration:	$_____	

9. Was the total project cost within maximum and minimum acceptable costs? If not, explain.
10. Was the total project duration within the maximum and minimum acceptable durations? If not, explain.
11. List unique lessons learned that might be of interest to upper management.

For instructions on how to set maximums and minimums of cost and duration, see Chapter 10, activity 2.

Item 8 indicates the maximum and minimum acceptable cost and project duration. To control the project, the project manager uses the upper and lower control limits of cost and schedule. These numbers are measured by the CPI and the SPI. Item 8 does not use these control limits to report final project achievement because upper managers are not familiar with the CPI and SPI concepts.

Figure 12.1   Final project evaluation outline. Use this outline to guide the final project evaluation. Use the results as part of the final report or briefing (Figure 12.2).

The content of the survey items may be stated as questions in an interview with the client or end users.

Estimate the time and budget needed to do this work package and include it in the WBS, budget, schedule, and resource plan, etc.

1. **Project Closeout Activities** (the information for this portion of the final report comes from Figure 12.1):
   (a) Statement that all work packages are complete.
   (b) Statement that the requirements of the business case definition and the charter have been fulfilled.
   (c) Statement that all contractual requirements with our customer have been met.
   (d) Statement that all contractual requirements with our subcontractors have been met.
   (e) Statement that all deliverables have been accepted.
   (f) Statement of project final evaluation data:
      Original estimate of project cost (OEPC)
      Total scope change budget
      Final project total budget
      Actual project final cost
      Cost variance at completion (CVAC)
      Cost of threat recovery
      Cost of opportunity recovery
      Total cost of risk recovery
      Used risk reserve as percentage of final project cost
   (g) CV: _____   Max. Acceptable Cost: _____
       SV: _____   Min. Acceptable Cost: _____
       CPI: _____   Max. Acceptable Duration: _____
       SPI: _____   Min. Acceptable Duration: _____
   (h) Was the total project cost within maximum and minimum acceptable costs? If not, explain.
   (i) Was the total project within maximum and minimum acceptable duration? If not, explain.
   (j) List unique lessons learned from the project.
2. **Client Closeout Activities** (the information for this portion of the final report comes from Figure 12.3):
   Customer Satisfaction Survey
   (a) Statement that all deliverables were accepted; if not, why not.
   (b) Feedback from customer satisfaction survey:
      How well the project was managed
      Quality of communications
      Customer concerns addressed
      Deliverables of acceptable quality
      Scope change procedure worked smoothly
      Project duration was reasonable
      Financial records clear and accurate
      Project cost was reasonable
      No outstanding issues
      Would use us again
   (c) Project manager's assessment of customer's propensity (likelihood) of giving us repeat business.

Figure 12.2 Outline for project final report or briefing. This outline can be used to write the project final report or briefing. The report would show the results of all the closeout activities. **(Continued on next page.)**

3. Organizational Closeout Activities:
   (a) Statement that project room was released to building management, and the date.
   (b) Statement that computers and associated equipment have been returned to the technical support department or to the rental company, and the date.
   (c) Statement that all financial records and funds have been reconciled to the satisfaction of the finance department.

4. Subcontractor Closeout Activities:
   (a) Statement that all subcontractor work and responsibilities have been completed.
   (b) Statement that all outstanding or expected invoices have been reconciled.
   (c) Statement that letters of appreciation to the subcontractors have been written.

5. Final Risk Assessment Activity:
   (a) Statement listing the threats identified as germane to the project closeout and the actions proposed.
   (b) Statement listing the opportunities identified as germane to the project closeout and the actions proposed.

6. Project Final Report: This outline is used to produce the final report.

7. Team Closeout Activities:
   (a) Statement that letters of appreciation have been written for all high-performing team members.
   (b) Statement that letters of appreciation have been written to functional managers for their resource support, their advice, and cooperation.
   (c) Statement that team members were released as of close of business (indicate the date).
   (d) Statement that the team held (or will hold) a celebration party to which all members of upper management were invited.

8. Report Closing Statement:
   (a) Statement of appreciation for the opportunity to lead this important project.
   (b) Statement of pride in submitting this report and in leading a great team of men and women.

Figure 12.2   (Continued).

Activity 3: Perform Organizational Closeout

The purpose of this work package is to conclude the team's use of organizational resources and to make a final resolution of the remaining resources. This work package consists of four subtasks as follows:

1. **Release the project room** by sending a memo to the facilities management office stating that the room will be vacated by a certain date, and available for other uses.

2. **Release computer equipment** by sending a memo to the technical support office (or rental company) indicating the day on which the equipment is available for other uses. In the memo, indicate the number of items that are being returned.

3. **Finalize financial records and funds** by preparing a finance resolution memo. In it, indicate any outstanding invoices against the project or invoices

Please use the following scale to answer the questions:

0	1	2	3	4	5	6	7	8	9	10
Strongly					Neutral					Strongly
Disagree										Agree

Use "NA" for not applicable. Please use comments to clarify or expand upon your rating

1. The project was well managed by the project team.   _____
   Comments:

2. Communications between the project team and us (me) were satisfactory.   _____
   Comments:

3. Our concerns were adequately addressed by the project team.   _____
   Comments:

4. The quality of the deliverables is acceptable.   _____
   Comments:

5. The scope change procedure worked smoothly.   _____
   Comments:

6. The duration of the project was reasonable.   _____
   Comments:

7. Financial records were clear and accurate.   _____
   Comments:

8. The cost of the project was reasonable.   _____
   Comments:

9. There are no outstanding issues yet to be resolved.   _____
   Comments:

10. If the opportunity arose, we would not hesitate to use your organization   _____
    (company) again.
    Comments:

Figure 12.3 Customer satisfaction survey.
Include the results of this survey in the final report (Figure 12.2).

---

expected, plus the balance of money that will remain after these invoices are paid. Figure 12.4 is an example of this memo.

4. **Prepare memos of appreciation** to functional managers and other stakeholders for their support in providing people, equipment, or advice. A sentence or two may pay big dividends in the future when you need assistance, advice, or resources from these same people.

Estimate the time and budget needed to do this work package and include it in the WBS, budget, schedule, and resource plan, etc.

Memo to: Finance Director
                     Ms Jane Powers
Memo from: Mary Iciak
Subject: Opus 12 Funds
Date: 11/28/07

The "Opus 12" project is concluded. All the work on the project has been completed including the two work packages performed by subcontractors (Ajax Corp. and Oliveri Corp.). The project room has been released to facilities management as of 12/1 and the computer rental company has been informed to pick up the computers and other rented equipment by 11/30. The rental invoice for November has been paid. Below is the final resolution of funds:

Project budget	1,200,500
Funds disbursed as of 11/30	1,100,000
Outstanding invoice (Ajax Corp.)	75,000
Expected invoice (Oliveri Corp.)	25,000
Team party (See invoice attached)	250
Balance	$250

Figure 12.4 Closeout memo to finance director.

Activity 4: Conduct Subcontractor Closeout

This work package consists of three subtasks:

1. **Determine whether subcontractors are finished:** Assure that all subcontractor responsibilities and work have been completed. Document this in a memo for record with a copy to finance so the subcontractors can be paid.
2. **Conduct final determination of monies due to the subcontractor firms:** Include this in the memo cited previously. You may need help from the procurement department.
3. **Prepare subcontractor letters of appreciation for each firm that has performed well:** Send a copy to procurement for their files.

Activity 5: Perform Risk Management Closeout

The purpose of this work package is to identify any threats or opportunities that are germane to the end of the project. Some of these threats and opportunities may have been previously identified and exist in the risk management plan. Some will be new and identified in this final risk assessment. The two subtasks to this work package are:

1. **Review the risk management plan:** Identify any threats and opportunities listed in the risk management plan that are pertinent to the closeout phase.
2. **Conduct a risk identification session:** To focus upon the threats and opportunities that are particular to the end of the project. Be aware that a

threat — something that might go wrong — could also be an opportunity for additional work and revenue.

Example 1: The project has developed a complex software application with procedures that users must follow carefully. The application is very powerful and efficient but only in the hands of experienced users. The project does not include training end users. The threat is that end users may not know how to use the software and blame the difficulties on the software design. If unaddressed, this situation constitutes a threat to the project team's reputation. This situation is also an opportunity for additional work for the customer by developing and conducting end-user training.

Figure 12.5 is a final risk assessment outline that can be used to identify threats and opportunities unique to the project closeout.

Estimate the time and budget needed to do this work package and include it in the WBS, budget, schedule, and resource plan, etc.

Activity 6: Write the Project Final Report

The purpose of the final report is to inform upper management of the completion of the project and to assure these people that all aspects of the project have been properly closed. You may get away without writing a final report, but your prestige as a project manager will be enhanced by a well-organized and professional final report. If necessary, the outline for the report may be used as the basis of a briefing. Figure 12.2 is an outline for the final report.

Activity 7: Conduct Team Closeout

The subtasks in this work package include:

1. **Conduct the final lessons learned session:** A lesson learned is an incident of experience that may be used to improve performance or repeat good performance. Something that the team did that worked should be retained as a lesson learned because the next time the team is faced with the same problem, it will want to use the action that was successful in the past. Something the team did that did not "work out" or something the team did not do that caused a subsequent problem need also to be retained as lessons learned. Figure 12.6 describes how to conduct a lessons learned session.

    Do not wait until the end of the project to conduct the lessons learned session. If you do so, many lessons will have been forgotten and the information will not help on the current project because the present project is concluding. A better strategy is to conduct this session after the team has

Residual risks

1. Are there any residual threats or opportunities in the risk management plan that can still impact the project ?

Deliverable(s) transference

2. Are there any threats or opportunities associated with the transference of the final deliverable(s) to the customer?

Improper operation

3. Is there a threat or opportunity that the functioning of the final deliverable(s) may be degraded by improper storage or use?

Training

4. Is there a threat or opportunity that end users do not have the training, experience, or attitude to use the final deliverable(s) properly and safely?

Maintenance

5. Is there a threat or opportunity that missed or improper maintenance of the final deliverable(s) may cause a failure or degraded functioning?

Fielding

6. Is there a threat or opportunity in the fielding (shipping, distribution, resupply, and repair support) of the final deliverable(s)?

Cash flow

7. is there a cash flow threat associated with getting paid?

Organizational acceptance

8. Is there a threat to the proper implementation of the project's output stemming from organizational politics? Are there any powerful stakeholders who want the project to fail?

Constituency acceptance

9. Are there any stakeholders or constituencies that could be useful in gaining wider acceptance od the project's deliverable(s)?

Legal Risks

10. Are there any legal threats to the conclusion of the project or the use of the project's deliverable(s)?

Figure 12.5   Final risk assessment.
Use this list of questions to identity threats and opportunities unique to the project closeout phase. Include the finding in the final report (Figure 12.2).

---

been together for about a month. Ask the same questions as indicated in Figure 12.6 but phrase them in the present tense (instead of "what went well?" use "what is going well?").

Use the lessons learned session every two or three months to get a sense of the team's morale, gather lessons learned, and shape behavior. The result of these sessions will be to highlight, reward, and clarify desired behavior. Rewarding behavior increases the probability of getting the same behavior again. The net result is that the team becomes more cohesive and committed to the project.

2. **Write letters of appreciation** to high-performing team members.
3. **Write letters of appreciation** to functional managers thanking them for their resource support, their advice and cooperation.
4. **Release team members** back to their functional departments.
5. **Celebrate** the project's conclusion with a party.

Estimate the time and budget needed to do this work package and include it in the WBS, budget, schedule, and resource plan, etc.

A lesson learned is a piece of experience that may be used to improve performance or repeat good performance. A lesson learned session is a team meeting in which the members share their perceptions of lessons learned. If possible, call in someone outside the team to serve as facilitator. The facilitator may use the brainstorming approach or the nominal group approach to generate the lessons learned. (Both of these techniques are discussed in Chapter 9, activity 2.)

The three questions to ask include:
What went well?
What didn't go well?
What should we have done differently?

The facilitator leads the brainstorming or nominal group session and records the ideas that come up. Focus upon one question at a time. When the team is finished surfacing ideas for all three questions, the facilitator may ask for clarification of some ideas.

Figure 12.6 Lessons learned small-group session.

Chapter 13

Execution Phase

Mega Recipe for the Execution Phase

Activity 1: Organize the team.
Activity 2: Coordinate and delegate work.
Activity 3: Establish performance control limits.
Activity 4: Purchase materials and services.
Activity 5: Monitor and control work packages.
Activity 6: Evaluate project progress.
Activity 7: Brief upper management and customer.
Activity 8: Deal with scope changes.
Activity 9: Monitor and control risks.
Activity 10: Revise planning and baseline documents.

Purposes of the Execution Phase

1. Produce the required deliverables.
2. Monitor work so the project's final cost and duration are within acceptable limits.
3. Deal with risk events.
4. Deal with scope changes.
5. Keep upper management or the customer informed of the project's status.
6. Achieve a high level of upper management or customer satisfaction with the team's efforts.
7. Document and disseminate lessons learned.

Role of the Project Manager

The project manager's role is to accomplish all the purposes of the project in a cost- and time-efficient manner while building and maintaining a relationship with upper management or the customer that will facilitate the acceptance and implementation of the team's deliverables. An additional role is to facilitate the recording and dissemination of lessons learned so future projects benefit from the current team's experiences.

Outputs of the Execution Phase

1. The accomplishment of all the deliverables at the appropriate cost and duration with satisfied upper managers and customers.
2. The successful application of the scope change procedure to control, document, and communicate the changes.
3. Acceptable cost of recovering from risks.
4. Forecasts of project final cost and duration times that were useful approximations.

Execution Phase Problems

It is during the execution phase that omissions, weaknesses, or errors from previous phases will be manifested. Some of the problems that occur during the execution phase include:

1. A constant stream of scope changes that seek to add, subtract, or otherwise modify the scope of the project. This is the result of a hasty initiation process with undiscovered requirements that become apparent during the execution phase. Too many scope changes can demoralize the team because they require the team to stop its momentum to investigate each scope change. Each scope change requires changes to the baseline documents and the plan. (Chapter 1 discusses how to avoid this problem.)
2. Many work package managers are not able to complete their work packages within acceptable cost and duration limits. Undisciplined estimating procedures are the cause. Failure to include a time or money reserve in the estimates for risky work packages is another cause. (Chapter 5, "Project Cost," and Chapter 6, "Project Schedule" have suggestions on how to avoid this problem.)
3. Upper management realizes that the project will require a much higher level of commitment (money, people, and time) than it originally understood. Because this commitment is not acceptable, upper management decides to discontinue the project. The fault lies in a truncated initiation phase where

the full extent of the organizational commitment should have been made clear. The very purpose of the business case definition briefing is to assure that upper management understands the problem the project will seek to solve and the organizational commitment required by the project before it made the decision to proceed. Another reason for this problem is a change in upper management people with the outgoing manager being replaced with someone who has different priorities for the organization. One strategy is to brief the incoming upper manager so the need for the project is understood and to build support if possible before the decision to discontinue the project is made. (See Chapter 2, "The Initiation Phase.")

4. An ongoing stream of changes and fixes are being made to the baseline and planning documents because of omissions and errors in the plan. This problem may have been the result of upper management pressure to get to the execution phase quickly "because we don't need to waste all that time on fancy, unnecessary planning documents." The reason may have been team inexperience. The team proceeded through the planning too quickly and is now having to make the fixes.

5. Earned value methodology (EVM), discussed in activity 6, requires an estimate of percent-complete for every work package. Inaccurate, distorted, or fanciful estimates will seriously jeopardize the credibility of the EVM status reports. The project manager has three strategies for overcoming this problem:

   a. Project manager discusses the method of determining percent of work accomplished with the work package manager at the delegation meeting. The project manager and the work package manager agree on the metrics that will be used to determine percent of work completed.

   b. An alternative strategy is for the project manager and work package manager to determine the percentage jointly.

   c. A third strategy is to assign the responsibility for estimating percent of work completed to a team member (management analyst) with a strong background in math.

Activity 1: Organize the Team

The execution phase requires the team to perform two kinds of work: completing work packages and providing functional support to the team manager. Managing the completion of work packages is the purpose of activities 2 through 5. Certain members of the team will need to be assigned the liaison role with the functional departments or specialized roles. Potential team member roles are listed as follows:

Liaison with procurement department: If the project includes the procurement of material and services, it will be necessary to assign a team member the responsibility to serve as liaison to the procurement department. It will be this person's responsibility to see that procured material or services arrive at the right time. This assignment should be given to a person with procurement experience — someone who knows the people in the procurement department and has a good working relationship with them. This person will also be responsible for monitoring procurement risks.

Schedule monitorship: This person should be very knowledgeable of the network diagram, will monitor it closely, and advise the project manager in schedule matters.

Earned value management: Someone on the team will need to be responsible for maintaining cost data plus earned value records and keep the project manager informed of the project's progress. This person must understand earned value technology and be experienced in spreadsheet applications.

Scope change point of contact: There needs to be a team member who is the point of contact and liaison for scope changes. It is this person who will manage the scope change procedure and perform the necessary liaison to determine the potential impacts of proposed changes.

Risk monitorship: Every team member should be assigned to monitor risk in the area of their specialty. The procurement representative will be assigned to monitor risks in the procurement area; the manufacturing member will be responsible for monitoring risks in the manufacturing area, etc.

Other areas that may require monitorship include finance, community relations, marketing, human resources, transportation, housing of team members, or physical and personnel security. Usually these responsibilities are ancillary and in addition to the other team member duties. Some of these roles may require full-time efforts when, for example, the project is being executed in a foreign country experiencing hostilities or lacking physical infrastructure such as roads, safe food and drinking water, medical care, and housing.

Activity 2: Coordinate and Delegate Work

A major portion of the project manager's efforts will be aimed at delegating and coordinating the work of others. Who is to perform each work package was decided when the resource plan (Chapter 7) was developed. However, the resource plan has information that may be months old, so it is always prudent for the project manager to do assertive, proactive coordination to assure "that the right person shows up at the right time, at the right place, with the proper understanding of the work package scope, with the materials and equipment needed to be successful." Assertive, proactive coordination means contacting (in the middle of May) the person who is to start a work package on June 9. Contacting the work package manager at least two weeks before the effort is to start gives the project manager the time to find a replacement work package manager if the previous one has been promoted, transferred, left the firm, etc. Finding out on the day a work package is to start that the work package manager is not available is a sure way to lose control of the schedule. To find another work package manager will take time.

The procurement liaison will keep the project manager informed about the timely delivery of materials or services.

A second major responsibility for the project manager during the execution phase is to delegate and initiate work in a way that assures that the performer has all the information with which to be successful. The work orders in Figure 4.1 (format) and Figure 13.1 (example) are designed to achieve this purpose.

Activity 3: Establish Performance Control Limits

Terminal baseline is the customer's expectation of what he will get (scope), when he will get it (duration time), and how much it will cost. For a project to hit its terminal baselines at the end of the project, it is necessary to establish the scope, cost, and duration time instrumental baselines for each work package and to communicate these to the work package managers. (See activity 3 in Chapter 10 and Table 10.1.)

Chapter 10 describes how maintaining work package performance within each package's time and cost control limits will assure achieving the customer's terminal baseline expectations when the project is concluded.

Activity 4: Purchase Materials and Services

The project manager needs to assign the responsibility for procurement to a team member. This person will use the information in the work package estimating sheet (Table 5.1 and Table 5.3) plus the information on the work order [Figure 13.1 (example) and Figure 4.1 (template)] to initiate the purchasing of materials and services. This person is responsible for writing the procurement memos requesting the materials and monitoring the process to assure that materials and services are purchased and available when needed — as specified in the resource plan (Table 7.1). The team's procurement liaison keeps the project manager informed of the status of procurements and any problems.

Activity 5: Monitor and Control Work Packages

The project manager's activities to establish control limits are to keep the total project cost and schedule variances within acceptable limits. Activity 3 through activity 5 in Chapter 10 describe how to establish the limits for each work package and how to communicate these requirements to the work package manager. The work order

Work Package Manager H. Green   **Work Package Title** Install hardwood floors   **Work Package No.** 4.6

Describe scope of the work package: Install ¾ inch prefinished oak flooring in all rooms on the first floor of the house except the bathroom, kitchen, and utility room. Install a vapor/noise sheet barrier under hardwood. Area is approximately 1,500 square feet. Materials supplied by general contractor.

                                                                              _____ **See Attachment**

Description of Deliverable(s): Floors installed as specified above.   _____ **See Attachment**

Budget Baseline $4,000   **Max Acceptable Cost** $4,444   **Min Acceptable Cost** $3,636 reference only
Effort Time Baseline Hours _____   (Days _____ × _____) (Weeks _____)
Max Acceptable Completion Time 12 Days   **Min Acceptable Completion Time** 8 days reference only
Early Start Date March 3, 2008   **Early Finish Date** March 14, 2008   *Total Float 2 Days
Latest Start Date March 5, 2008   **Latest Finish Date** March 16, 2008
* None of the total float may be used without the Project Manager's prior approval.

Potential Problems or Risks
    Temperature inside the house must be at least 50 degrees F.

Acceptance Criteria (Describe how performance will be evaluated)   **Other (Describe):** Floor has no blemishes and no spaces between boards of more than ⅛ of an inch.
    _x_ Acceptable Deliverables   _____ **See Attachment**
    _x_ Cost with Max and Min Control Limits
    _x_ Performance Time within Max and Min Control Limits

Figure 13.1 Work order example. (The blank form is shown in Figure 4.1)

(See Figure 4.1 template and Figure 13.1 example.) is used to document and communicate this information. The control matrices in Table 10.1 (example) and Table 10.2 (template) show how to create a matrix of upper and lower limits for all work packages. Figure 10.1 is an example of an Excel®-generated matrix and Figure 10.2 is the Excel spreadsheet that will do all the calculations and produce the matrix.

When delegating work, the project manager needs to encourage work package managers emphatically to keep performance within the established cost and schedule limits. The project manager needs to monitor performance (get feedback) at the 25% mark for each work package. It would be imprudent to get feedback about a work package in the middle of the second week for a work package that has a 10-day baseline duration. Finding out that the work package manager is having difficulty two days before the work package is scheduled to complete is too late to intervene and correct the problem. The longer the project manager waits, the more difficult the correction will be to accomplish. Monitor work package performance early and frequently. For a 10-day work package, start monitoring performance at the middle of the first week.

Activity 6: Evaluate Project Progress

The **earned value methodology (EVM)**[1] (equations) is used to evaluate project progress. To be able to use this technology, the project manager must have three pieces of information for every work package: planned value (PV), earned value (EV), and actual cost (AC).

Planned value (PV) is the original budget for a work package and it is the dollar value of work scheduled to be completed by the data date. The data date is the day on which the project's progress is being evaluated. For example, a work package may have a total budget of $2,000 and half this amount ($1,000) is scheduled to be completed by the data date. If the entire work package is scheduled to be complete by the data date, then the total planned value (PV) and the planned value as of the data date (PV_{dd}) are the same. On the project Gantt chart (see Figure 13.2, work package E), these two PVs are different, the total PV is $10,000, and the PV_{dd} is $5,000. Only for those work packages split by the data date or starting after the data date (work package F in Figure 13.2) will have a PV and PV_{dd} that are different numbers. Otherwise, the two PVs are the same number.

> **Example 1:** A work package is estimated to take 40 hours to complete and it will require a software developer that has a $50 per hour loaded labor rate. The department for which this developer works charges $50 for every hour of this person's time. There are no material or equipment costs. The estimate equals $2,000 (40 hours × $50 per hour). Once approved, the estimate becomes the budget, the PV, for this work package.

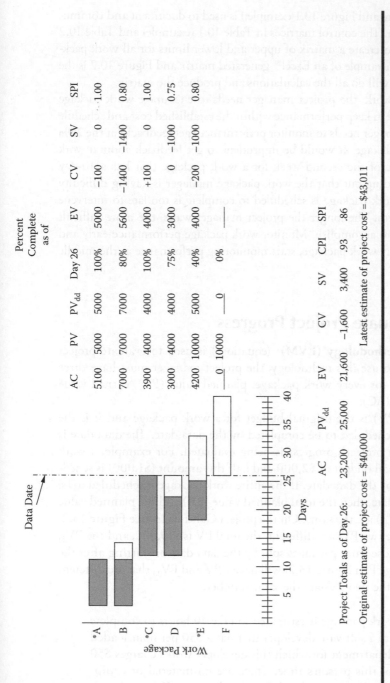

Figure 13.2 Project evaluation using earned value methodology. *Work packages A, C, and E are on the critical path. The CPI was omitted from this figure because of lack of space. Work packages A, B, and E are over budget by $100, $1,400, and $200, respectively. Work package C is under budget by $100 and D is right on budget. Work packages A and C are on schedule because their schedule variances are zero and their SPIs are 1.00. Work package B is $1,400 dollars worth of work behind schedule; D and E are both $1,000 worth of work behind schedule. The SPIs for B, D, and E are all less than 1.00, meaning behind schedule. Work package D was scheduled to complete in 14 days; nine days have been expended. Equation 13.6 determines that it is 2.25 days behind schedule (.75 × 9 days − 9 days = −2.25 days). The project is in day 26 and scheduled to complete on day 40. Equation 10.11 is LEPD = OEPD ÷ SPI$_{cp}$. The SPI of the critical path (SPI$_{cp}$) equals the sum of the EVs on the critical path ÷ sum of the PV$_{dd}$ on the critical path. Work packages A, C, and E are on the critical path. EV$_{cp}$ = 5,000 + 4,000 + 4,000 = $13,000; PV$_{cp}$ = 5,000 + 4,000 + 5,000 = $14,000; SPI$_{cp}$ = EV$_{cp}$ ÷ PV$_{cp}$ = $13,000 ÷ $14,000 = .93. Original estimate of project duration (OEPD) is 40 days. Latest estimate of project duration (LEPD) = OEPD ÷ SPI$_{cp}$ = 40 days ÷ .93 = 43 days. Status of the project as of day 26: $1,600 over budget and $3,400 behind schedule. Total project CPI = $21,600 ÷ $23,200 = .93. Final cost of the project is estimated at $43,011 ($40,000 ÷ .93). Figure 10.3 is an example of an earned value report developed by an Excel® spreadsheet. Figure 10.4 is a blank spreadsheet, which includes the embedded earned value equations.

Earned value is the dollar value of work completed. The formula for earned value is shown as follows:

$$\text{Earned value} = \text{percent complete} \times \text{total planned value} \quad (13.1)$$

$$\text{EV} = \% \text{ complete} \times \text{total PV}$$

Example 1 continued: The work package is scheduled to be completed today. It was scheduled to take five days to complete. The developer tells the project manager that the work package will be 90% completed by the end of day 5. The earned value is $1,800 (.90 × 2,000).

Actual cost is the dollar amount charged for completing the work package. This number comes from a financial bill-out sheet that lists all ongoing work packages and the charges against each. For example 1, the actual cost (charges) for the work package is $1,900.

With these three pieces of information we can use the following earned value equations:

$$\text{Cost variance} = \text{earned value} - \text{actual cost} \quad (13.2)$$

$$\text{CV} = \text{EV} - \text{AC}$$

$$\text{Schedule variance} = \text{earned value} - \text{planned value as of the data date} \quad (13.3)$$

$$\text{SV} = \text{EV} - \text{PV}_{dd}$$

When using Equations 13.3 and 13.5, interpret PV_{dd} as that part of (dollar amount) the total PV scheduled to be completed by the data date.

$$\text{Cost performance index} = \text{earned value} \div \text{actual cost} \quad (13.4)$$

$$\text{CPI} = \text{EV} \div \text{AC}$$

$$\text{Schedule performance index} = \text{earned value} \div \text{planned value as of the data date } (\text{PV}_{dd})$$

$$(13.5)$$

$$\text{SPI} = \text{EV} \div \text{PV}_{dd}$$

Example 1 continued: The status of this work package is as follows:

$\text{CV} = \text{EV} - \text{AC}$	$\text{CV} = \$1,800 - \$1,900 = -\$100$
$\text{SV} = \text{EV} - \text{PV}_{dd}$	$\text{SV} = \$1,800 - \$2,000 = -\$200$

PV_{dd} is \$2,000 because all of the work package, i.e., \$2,000 worth of work was scheduled to be completed by the data date.

CPI = EV ÷ AC	CPI = \$1,800 ÷ \$1,900 = .95
SPI = EV ÷ PV_{dd}	SPI = \$1,800 ÷ \$2,000 = .90

Example 2: A work package has a total budget (PV) of \$5,000. It is scheduled to take 10 days to do; start on day 10 and end on day 19. An evaluation of work package progress is made on day 15 when only half of the work package was scheduled to be complete. On day 15, 40% of the work package PV is completed. The total budget (PV) for the work package is \$5,000 but the amount of work scheduled to be completed by day 15 is \$2,500. The total PV = \$5,000 and the PV_{dd} = \$2,500. The earned value = % complete × total PV = .40 × \$5000 = \$2,000. The schedule variance = EV − PV_{dd} = \$2,000 − \$2,500 = −\$500. The work package is \$500 worth of work behind schedule.

The earned value equations can be used to determine the status of a project as well as the work packages in the project. Figure 13.2 shows the Gantt chart for a simple project consisting of six work packages. It shows the earned value status of the total project as well as that for each work package.

How to interpret earned value data: Positive variance is good and negative variance is bad. Cost performance and schedule performance index numbers less than one are bad, and index numbers greater than one are good. There are only three conditions that apply to a cost variance: over budget, under budget, or right on the budget. The cost status of a work package must be one of these three. A minus cost variance is bad, which means over budget; a positive cost variance is good, which means under budget; and a zero cost variance is perfect because it means right on budget. Schedule variance is interpreted the same way; the three conditions of schedule variance are behind schedule, ahead of schedule, or right on schedule. A negative schedule variance is bad, and means behind schedule. A positive schedule variance is good, and means ahead of schedule. A zero schedule variance is perfect, and means right on the schedule.

The cost performance index (CPI) and the schedule performance index (SPI) are both measures of efficiency. The CPI measures the efficiency with which money is used. A CPI of 1.05 is good because it indicates that the project has earned \$1.05 in value for every dollar spent. A CPI of .85 is bad because it indicates that the project has earned \$.85 for every dollar spent. A CPI of 1.00 means that for every dollar spent, one dollar in value was achieved. Thus a CPI of 1.05 means the team is 105% efficient in the use of money; this suggests that the final cost will be less than budgeted. A CPI of .85 means the team is 85% efficient in the use of money and will need more than the original budget to complete the project.

The SPI is interpreted in the same way as the CPI. An SPI of 1.20 is good because it indicates that the team is achieving 120% of a day's work for every day expended. The project is ahead of schedule. An SPI of .90 is bad, and indicates that the project team is achieving 90% of a day's work for every day expended. The project is behind schedule. A schedule performance index of 1.00 means that the project is exactly on schedule.

To determine the days the work package or project is ahead or behind schedule, use the following equation:

Days ahead/behind schedule = (SPI × days consumed) − days consumed

(13.6)

Example 3: A work package had consumed 10 days of effort and its SPI is .90. Days behind schedule = .90 × 10 days − 10 days = 9 days − 10 days = −1 day. The negative sign in front of the 1 indicates that it is one day behind schedule.

Example 4: A project has been ongoing for 40 days and its SPI at day 40 is 1.20. Days ahead of schedule = 1.20 × 40 days − 40 days = 48 days − 40 days or 8 days ahead of schedule. The plus sign in front of the 8 indicates ahead of schedule.

The CPI and the SPI may be used to forecast the project's final cost and duration time, respectively. Forecasting final project cost is Equation 10.12 (repeated below) and forecasting final duration time is Equation 10.11 (repeated below).

Latest estimate of project cost (LEPC) = Original estimate of project cost (OEPC) ÷ cost performance index (CPI) (10.12)

$$\text{LEPC} = \text{OEPC} \div \text{CPI}$$

Example 5: Figure 13.2 describes a project with an original budget of $40,000 and a CPI at day 26 of .93. The latest estimate of project cost = $43,011 (40,000 ÷ .93).

Latest estimate of project duration (LEPD) = original estimate of project duration (OEPD) ÷ schedule performance index along the critical path (SPI_{cp}) (10.11)

$$\text{LEPD} = \text{OEPD} \div \text{SPI}_{cp}$$

Example 6: Figure 13.2 indicates a project with an original estimate of project duration (OEPD) of 40 days. At day 26, the schedule performance index along the critical path (SPI$_{cp}$) is estimated at .93. The latest estimate of project duration (LEPD) = 43 days (40 days ÷ .93).

Equation 10.11 requires the SPI only along the critical path. It does not use the SPI for all work packages because only the work packages on the critical path determine the project's duration. Determine the earned values (EV) for all work packages on the critical path; determine their sum. Note the planned value (PV) for all work packages on the critical path; determine their sum. The equation for the SPI along the critical path is

SPI on critical path (SPI$_{cp}$) = total earned value on the critical path (EV$_{cp}$)
÷ total planned value on the critical path (PV$_{cp}$)

(10.10)

$$SPI_{cp} = EV_{cp} \div PV_{cp}$$

Example 7: Figure 13.2 indicates a project with three work packages on the critical path: A, C, and E. The total earned value (EV$_{cp}$) for these three work packages is $13,000. The total planned value (PV$_{cp}$) is $14,000. The schedule performance index along the critical path (SPI$_{cp}$) = .93 (13,000 ÷ 14,000).

The status of the work package in example 1 is as follows. Cost variance (CV) of negative $100 means it is $100 over budget. The schedule variance of negative $200 means it is $200 worth of work behind schedule. The CPI of .95 indicates that it is over budget and will need more than the original budget to complete. The SPI of .90 indicates it is behind schedule. The work package was scheduled for five days. Equation 13.6 indicates that it is .9 × 5 days − 5 days = −.5 days; the minus indicates that the work package is one-half day late.

Figure 13.2 is an example of using the earned value methodology to evaluate a project's progress. It also demonstrates how to determine the SPI for the critical path, project final cost, and the latest estimate of duration time.

Activity 7: Brief Upper Management and Customer

There are three times when upper management should be briefed:

1. Shortly after the start of the execution phase. The purpose of this briefing is to communicate that the project has been successfully launched and is ongoing. This briefing includes the planning documents, so upper managers may assess how prepared the team is to execute the project and to judge the feasibility of the plan. Include the following topics:

    a. Need for the project
    b. Purpose of the project
    c. Planned deliverables
    d. Work breakdown structure
    e. Budget
    f. Schedule (Gantt chart)
    g. Major potential threats and suggested strategies
    h. Major potential opportunities and suggested strategies
    i. Size of the reserve
    j. Resource requirements (resource plan)
    k. Outline of the scope change procedure
    l. The communication plan

2. The midpoint of the execution phase is the second time a briefing is needed. The purpose of this interim briefing is to present the status of the project and forecast final project cost and duration. Include the following topics:
    a. Total project cost and schedule variance with explanations if necessary
    b. Cost and schedule performance indices with explanations if necessary
    c. Gantt chart showing project progress
    d. Risk events that have occurred, the recovery strategies used, and the effectiveness of the strategies
    e. Balance of reserve
    f. Balance of budget
    g. Percent of work completed (Equation 13.7)
    h. Percent of money spent (Equation 13.8)
    i. Forecast of total project duration (Equation 10.11)
    j. Forecast of total project cost (Equation 10.12)
    k. Strategies for correcting the over budget or lateness problems indicated in (i) and (j) if necessary

If the project is a very long one or one with high visibility, interest, or cost, it may be necessary to have more than one interim briefing.

Percent of work complete[2] = total project EV as of the data date
$$\div \text{ original estimate of project cost (OEPC)} \quad (13.7)$$

$$\% \text{ work complete} = EV_{dd} \div OEPC$$

Example 8: Figure 13.2 shows a project with an original budget of $40,000. The total earned value as of day 26 (EV_{dd}) is $21,600. The percent of work complete is .54 (21,600 ÷ 40,000).

Percent of money spent[2] = total project AC as of the data date
$$\div \text{ original estimate of project cost (OEPC)} \quad (13.8)$$

$$\% \text{ money spent} = AC_{dd} \div OEPD$$

Example 9: Figure 13.2 indicates the project has spent $23,200 as of the data date. The original project budget is $40,000. The percent of money spent = .58 (23,200 ÷ 40,000).

3. The third time a briefing is needed is at the end of the project when the purpose is to communicate that the customer is satisfied and all aspects of the project have been completed. The outline for the final report, Figure 12.2, lists the topics and information that are included in this last briefing.

Activity 8: Deal with Scope Changes

Chapter 11 discusses the scope change procedure and the forms that should be used. Assign one team member the responsibility to facilitate the scope change process.

Activity 9: Monitor and Control Risks

Two kinds of threats need to be monitored and controlled: the extraordinary threats listed in the risk management plan and the many incidental threats.

For the extraordinary threats listed in the risk management plan, the project manager's role is to assure that every team member recognizes the area of risk he or she is responsible for monitoring. This is discussed in Chapter 9, activity 1. Prepare the team to do risk management. Each team member will monitor one area of risk and note any situations, circumstances, or attitudes that may affect the probability, impact, timing, or strategy effectiveness. Any changes in these should be brought to the project manager's attention.

In addition to monitoring the status of identified threats, each team member is also responsible to gauge the changing status of each opportunity. Changes in situations, circumstances, or attitudes that may affect probability, impact (benefit), timing of opportunities, or strategy effectiveness must be brought to the project manager's attention.

The team implements the strategy in the risk management plan for any identified risks, threats, or opportunities that occur. The effectiveness of the strategy will be assessed. Strategies perceived as successful will be retained and continued. Those judged to be unsuccessful will be replaced with another strategy.

The responsibility of dealing with the incidental threats rests with the project manager. Incidental threats are those circumstances, situations, or attitudes that prevent the fruition of the following sentence: "The right person shows up at the right place, at the right time, with the complete understanding of the work package scope, and the materials, equipment, etc., with which to be successful." A work package manager not showing up for the start of a work package, the absence of the needed materials or equipment, or lacking information can all result in lost time and money. The project manager uses the resource plan, Table 7.1, to do advanced assertive coordination. The purpose of this activity is to assure that all the resources are available where and when needed to execute the project. Failure to do this will result in many negative impacts to the project cost and schedule.

Activity 10: Revise Planning and Baseline Documents

The scope change procedure (Chapter 11) requires changing the planning documents and baseline documents for every approved change. Changes and improvements to the plan directed by the project manager also need to be incorporated into the plan and baseline. The project manager needs to assign a team member to perform this duty.

Notes

1. *Project Management Body of Knowledge* (PMBOK®), Third Edition, 2004. Earned value equations: CV, SV, CPI, SPI, and EAC, pp. 173–176. Project Management Institute, 4 Campus Boulevard, Newtown Square, PA 19073. With permission.
2. The source for these equations is *Scheduling and Cost Control Instructor Guide*, page 6–15, July 2005. © ESI International, 901 North Glebe Road, Suite 200, Arlington, VA 22202. With permission.
3. Estimate at completion (EAC) is an alternative for the preferred term, latent estimate of project cost (LEPC). Budget at completion (BAC) is an alternative for the preferred term, original estimate of project cost (OEPC).

Chapter 14

Closeout Phase

Mega Recipe for Closing the Project

The purpose of the project closeout activities is to end the project in a way that reflects favorably upon the team, the team leader, and the organization. This phase requires the completion of the seven activities planned in Chapter 12. It is fortuitous that comprehensive planning makes the closeout phase rather straightforward.

> **Activity 1:** Perform project closeout.
> **Activity 2:** Perform client closeout.
> **Activity 3:** Perform organizational closeout.
> **Activity 4:** Conduct subcontractor closeout.
> **Activity 5:** Perform final risk assessment.
> **Activity 6:** Write project final report or briefing.
> **Activity 7:** Conduct team closeout.

Purpose of the Closeout Phase

The purpose of the closeout phase is to conclude all facets of the project to the satisfaction of upper management before the team members start to leave the team. Treating the seven activities listed above as ordinary work packages makes this phase a continuation of the execution phase.

Role of the Project Manager

The project manager's role in the closeout phase is to assure that all aspects of the project are properly concluded. At the end of the project, team members often lose

233

their focus; they lose some of their discipline for showing up on time, for concentrating on the task at hand, etc. It is the project manager's responsibility to help the team retain its focus during the closeout activities.

Outputs of the Closeout Phase

Each of the seven activities will produce a memo to the project manager stating the results of the activity. The capstone output of this phase is the project final report, written by the project manager, which summarizes the results of the seven activities.

Closeout Phase Problems

1. The team failed to plan out the closeout phase and is attempting to plan and execute at the same time. The resources of people, money, and time have not been planned, so it is a haphazard effort at best with key closeout activities being missed or truncated. The solution is to plan the closeout activities at the same time the rest of the work breakdown structure (WBS) is developed. Include the closeout work packages (activities) in the budget, schedule, WBS, etc. Include the closeout activities in the resource plan so each team member knows his or her responsibilities during closeout.

2. Team members start to leave the team before the closeout phase is completed because they do not understand all that has to be done and their roles in the closeout. The project manager should tell the team early in the project that the closeout phase consists of activities that include all team members. The project manager should build a sense of team cohesion so individual members do not consider leaving the team early.

3. Functional managers start to withdraw their team members before the closeout is complete. Solution: In the interim briefing to upper management, the project manager indicates that the project closeout consists of a number of important closeout activities that will require the full team to perform — with the request that functional managers not take team members back until released by the project manager.

4. A potential problem is when the team surfaces a number of unresolved issues, uncompleted work packages, or unacceptable deliverables. This means the project is not finished. The project manager will have to remain on the team (perhaps with a small staff) after the team disbands to resolve the problems. One preventive solution: The project manager communicates frequently with the customer especially when it is time to get the customer's approval (in writing if possible) of work performed, deliverables submitted,

or problems resolved. Don't wait until the end of the project to resolve issues in the hope that they will have gone away by then. A memo for the record is a good way to document that something happened: a deliverable submitted to the customer because it does not require the customer to sign anything. If it is inappropriate to ask a customer, a stakeholder, or upper manager to sign a document, use the memo for record. In the absence of a negative response from the upper manager, we conclude the action, deliverable, etc., is acceptable. A statement of the following kind might help clarify this: "In the absence of any feedback within the next 10 business days, we will conclude that this action [deliverable, etc.] is acceptable."

Read Chapter 12 and the project's closeout plan before commencing the closeout activities.

Activity 1: Perform Project Closeout

The purpose of this work package is to determine that all the requirements have been satisfied. Assign one or two team members to perform the following subtasks:

1. Validate and document that all the work packages listed on the WBS have been completed. If any work package is not complete for some legitimate reason, document the reason.
2. Review the business case definition and charter to assure and document that all requirements on these documents have been completed or resolved.
3. Review the contract, if there is one, to determine that all requirements have been met.
4. Conduct the final project evaluation. Use the outline in Figure 12.1 as a guide in analyzing and documenting the results.

Each subtask should be included in the memo to the project manager indicating the findings and conclusions. Memos should be short, simple, and direct. Figure 14.1 is an example.

Activity 2: Perform Client Closeout

1. Assure that the client has accepted the deliverables. This requires someone to interview the client and determine that all deliverables have been received and accepted. A memo to the project manager is how this subtask is documented.
2. Measure the degree to which the client is satisfied. This is greatly facilitated by using the customer satisfaction survey in Figure 12.3. The question of whom to survey is important.

Memo to:	John Albertson,
	Opus 12
	Project Manager
From:	Helen Coulieri,
	Team Member
Subject:	Project Closeout Activity 1
Date:	Aug 15, 2008

Tom Johnson and I have completed work package 14.1 Perform Project Closeout with its four subtasks and have the following results to report:

(a) Validate project completion. A review of the WBS revealed that all work packages have been completed. Attached is the "final WBS," which we have annotated to indicate that each work package is complete.

(b) Review the business case definition and charter. A review of these documents indicates that all commitments therein have been met. An interview with our sponsor indicated that he was satisfied that all requirements had been met. Attached are the record copies of these annotated documents.

(c) Review the contract. A review of the contract revealed one requirement that had not been completed. The contract requires that our company certify that the project has been an equal opportunity effort. Attached is a letter for your signature to that effect. All other requirements have been met.

(d) Conduct a final project evaluation. We used the outline in Figure 12.1 to conduct the final project evaluation. The results of our findings are indicated on the final project evaluation report attached.

Incl: as

Figure 14.1   Project closeout memo.

The output of this work package is a memo to the project manager indicating the results of each subtask. Figure 14.2 is an example.

Activity 3: Perform Organizational Closeout

The aim of this work package is to conclude the team's use of organizational resources and to make a final resolution of the remaining resources. Select a team member to complete the four subtasks in this work package.

1. The first subtask requires the writing of a memo to the facilities management office stating that the project room will be vacated on a certain date. A copy of this memo is sent to the project manager along with the report on the entire work package.

2. Releasing all borrowed or rented equipment is the second subtask.(The team member will write a memo to the technical support office if the equipment had been borrowed from that office or a letter or telephone call followed up by a letter to the rental company from whom the equipment was rented.) The memo or letter will need to include an inventory of equipment

Memo to: John Albertson,
                        Opus 12
                        Project Manager
From: Henre La Poire,
                        Team Member
Subject: Client Closeout Activity 2
Date: Aug 19, 2008

Work package 14.2 is complete. The results of subtasks (a) and (b) are shown below:

(a) On Aug 14, 2008 I spoke to Paul Pazzo, the client's contracting officer technical representative (COTR). He affirmed verbally that they have received all the deliverables noted in our contract and they are acceptable. I have written a memo to Paul (dated 12 Aug 08) confirming our conversation and his statement that all deliverables are acceptable. It is attached.

(b) I used our customer satisfaction survey (Figure 12.3) to get feedback from five people in the client organization: Mary Smart, VP of Finance; Carmen Giordino, Division Manager of Manufacturing; Janette Hoener, Procurement Division Chief; Paul Pazzo, COTR; and William Castleborn, the software engineering chief. The last two people are the ones we worked with on an-almost daily basis.

These people were kind enough to sign their surveys so we can determine where we succeeded and where we didn't. The VP of Finance is one tough person; her scores were always lower than those of the others. The surveys are attached.

In general, the surveys are quite positive. On a scale of 0 (strongly disagree) to 10 (strongly agree) we got scores as follows:

Project well managed	Average 8.5
Communication was satisfactory	Average 10.0
Their concerns addressed	Average 9.3
Deliverables are acceptable	Average 9.1
Scope change procedure worked smoothly	Average 9.6
Financial records clear and accurate	Average 8.0
Cost reasonable	Average 6.9
Outstanding issues resolved	Average 9.1
Would use us again	Average 9.2

The only negative comment came from the VP of Finance, who stated that the project was more expensive than expected although the charges seem valid.

Our team got two favorable comments about how smoothly the scope change procedure worked. Someone did a great job communicating (schmoozing) because all respondees rated this as a 10.

One comment said, "Would use them again if we can get the price down!" This came from the VP of Finance.

Incl: as

Figure 14.2   Client closeout memo.

---

being returned. This subtask includes reconciling any differences between the original inventory and the return equipment inventory. A copy of the closure memo will be provided to the project manager.

3.   Finalizing financial records and funds is the third subtask in this work package. The person doing this subtask will reconcile the original project budget against monies spent, outstanding invoices, and expected incoming

Memo to: Ms. Janette Hoener,
                    Chief, Procurement Division
From: John Albertson,
                    Opus 12
                    Project Manager

Subject: Letter of Appreciation
Date: Aug 12, 2008

Dear Ms. Hoener,

    I would like to thank you and your professional staff for their great support in the procurement of our project's materials and services. I personally learned a lot about procurement from your agents because they were so willing to explain the process and nuances. I am confident that the project would not have progressed as smoothly as it did without the extra effort provided by the Procurement Division.

Please convey my personal appreciation as well as that of the team.

Figure 14.3   Example letter of appreciation.

---

    invoices plus the balance that will remain after all invoices are paid. The process and documentation must be acceptable to the finance department; the financial records are not reconciled until the finance department concurs. Figure 12.4 is an example of a closeout memo to the finance director.

4. Preparing memos of appreciation to functional managers and other stakeholders is also a subtask of this work package. In each memo, try to identify a unique contribution or help received for which the team is grateful. Figure 14.3 is an example. Prepare these memos for the project manager's signature.

    Write a memo to the project manager stating the results of activity 3. Figure 14.4 is an example.

Activity 4: Conduct Subcontractor Closeout

The purpose of this work package is to reconcile all subcontractor matters:

1. Determine that the subcontractors have completed all the work for which they are responsible. Include the results of this investigation in the memo to the project manager.
2. Reconcile the amount of monies due the subcontractors. Coordinate with the procurement department on this. Include the results of this reconciliation in the memo to the project manager.
3. Prepare subcontractor letters of appreciation for each firm that performed well. Write the letters for the project manager's signature and send a copy to the procurement department.

    Write a memo to the project manager stating the results of activity 4. Figure 14.5 is an example.

Memo to:	John Albertson,
	Opus 12
	Project Manager
From:	Mary Iciak,
	Team Member
Subject:	Organization Closeout Activity 3
Date:	Aug 19, 2008

Work package 14.3 was completed today. The results of the four subtasks are listed below:

(a) I contacted the Facilities Management Office and informed them that the team will be vacating the office space on Oct 31st. I also sent a confirming memo, a copy of which is attached.

(b) I contacted the Technical Support Office and told them to pick up the computers and associated equipment by noontime on Oct 31st. I have prepared a final inventory, which corresponds with the list of equipment we received at the beginning of the project. A copy of the final inventory will be signed by someone from tech support and I will give it to you on Oct 31. I have contacted the company that rented us the photocopy machine. They will pick it up by noon on Oct 31. I'll make sure we get a receipt. The Facilities Management Office said they will not remove the telephones because the next tenants will need them anyway.

(c) I reconciled the Opus 12 budget against expenditures. The reconciliation memo to the Finance Director is attached.

(d) I have prepared letters of appreciation for the functional managers, as we discussed last week. They are attached and ready for your signature.

Figure 14.4 Organization closeout memo. See Figure 12.4 for an example of the memo to the finance director. See Figure 14.3 for an example of a letter of appreciation.

Memo to:	John Albertson,
	Opus 12
	Project Manager
From:	Robert Mungombo,
	Team Member
Subject:	Subcontractor Closeout Activity 4
Date:	Aug 18, 2008

Today I completed work package 14.4. The results of the subtasks (a), (b), and (c) are listed below:

(a) I have reviewed our contracts with each of the two subcontractors (Ajax Corp. and Oliveri Corp.) as well as discussed the subcontractor performance with the COTR Paul Pazzo, who confirmed that they have performed all the required work, produced the necessary deliverables, and all are acceptable.

(b) I contacted the Procurement Department, which informed me that the Ajax Corp. invoice for $75,000 has been received. The Oliveri Corp. invoice for $25,000 has not been received. The Procurement Department has approved the invoice amounts listed above. There are no other subcontractor invoices expected.

(c) I have prepared a letter of appreciation to each of the subcontractors for your signature. They are attached.

Figure 14.5 Subcontractor closeout memo.

Activity 5: Perform Final Risk Assessment

The purpose of this work package is to identify any threats or opportunities that are relevant to the end of the project. This work package will require a small group to perform because it involves performing activities 2 through 5 in Chapter 9:

> Identify risks
> Estimate probability and impact
> Stratify risks
> Develop strategies

Use the final risk assessment outline (Figure 12.5) as a guide for identifying risks. It lists ten categories of risks that should be investigated and discussed as follows:

1. Residual risks
2. Deliverable transference
3. Improper operation
4. Training
5. Maintenance
6. Fielding
7. Cash flow
8. Organizational politics
9. Constituency acceptance
10. Legal risks

Use a small-group generating process (see Chapter 9, activity 2) to identify threats and opportunities under each of the ten categories.

The following formats will be useful: Table 9.1 and Table 9.2 formats for threat list and opportunity list, respectively; Table 9.11 format for threat strategies list; and Table 9.12 format for threat and opportunity strategies interaction.

This work package requires a final risk assessment report, which is presented to the project manager for discussion and dissemination. The report consists of the documents listed above plus a cover memo summarizing the findings. Figure 14.6 is an example of a final risk assessment report.

Activity 6: Write the Project Final Report

The final report summarizes all the closeout activities. Its purposes are to demonstrate to upper management that the project is indeed complete, and to present the final status of the project. The report will include the following topics:

Memo to: John Albertson,
                Opus 12
                Project Manager
From: Manny Maple
Subject: Final Risk Assessment

The team of Helen Troy, Janet Schoener, and I have completed the final risk assessment for the Opus 12 project. We considered the following categories of risks: (a) maintenance, (b) fielding of the system, (c) case flow, (d) organizational politics, (e) constituency acceptance, (f) legal, (g) deliverable transference, (h) improper operation of the system, and (i) training. Here are our findings:

1. **Maintenance.** We talked to the people who will actually use our system and found out that our client is notoriously bad in committing resources (actually not committing resources) for maintenance mainly because the client doesn't have the trained people to do it. This is a threat to our reputation because the system will fail if it is not properly maintained. It is an opportunity for additional work — either we do the maintenance or train them on how to do it.

2. **Organizational politics and constituency acceptance.** Our client has a centralized headquarters with semi-autonomous regional offices. The regional offices are not convinced that they need our system. Without the regions using our system, they will come to think of our work as a useless waste of money. It's key that the regions be convinced of the value of our system for them. If the regions do not adopt our system, this entire project will eventually be conceived as a folly. This is a threat wrapped around an opportunity. If we could demonstrate the utility of our system in the regions (with headquarters involvement), we would save the credibility of our system but also gain entry into the regions which have the authority to develop their own independent systems.

3. **Residual risks.** All of the risks in the risk management plan are either resolved or behind us. No further action is necessary for those risks in the risk management plan.

4. **Training.** There are two opportunities for training: (1) we could conduct user training at the regions as our strategy of showing the power and ease of our system. We wouldn't have to train all the users at the regions, just a few influential users who could convince the others. We shouldn't charge them for this initial training but charge them for the subsequent training. (2) There is also an opportunity to train the headquarters and regional personnel in system maintenance. The least we should propose is a maintenance manual.

Figure 14.6   Final risk assessment report.

---

1. Project closeout activities
2. Client closeout activities
3. Organizational closeout activities
4. Subcontractor closeout activities
5. Final risk assessment
6. Team closeout activities
7. The report closing statement

The person writing the report has the closeout memos for the activities. For examples of closeout memos, see Figure 14.1 (project closeout), Figure 14.2 (client closeout), Figure 14.4 (organization closeout), and Figure 14.5 (subcontractor closeout). Figure 12.2 is an outline of the final report.

Activity 7: Conduct Team Closeout

1. **Conduct the final lessons learned session:** A lesson learned is a piece of experience that may be used to improve performance or repeat good performance. Something that the team did that worked well should be retained as a lesson learned because the next time the team is faced with a similar problem, it will want to use the action that was successful in the past. Something the team did that did not "work out" or something the team did not do that caused a subsequent problem need to be retained as lessons learned. Figure 12.6 describes how to conduct a lessons learned session.

   Do not wait until the end of the project to conduct the lessons learned session. If you do so, many lessons will have been forgotten and the information will not be available to help on the current project because the present project is concluding. A better strategy is to conduct this session after the team has been together for about a month. Ask the same questions as indicated in Figure 12.6 but phrase them in the present tense (instead of "what went well?" use "what is going well?").

   Use the lessons learned session every two or three months to get a sense of the team's morale, to gather lessons learned, and to shape behavior. The result of these sessions will be to highlight, reward, and clarify desired behavior. Rewarding behavior increases the probability of getting the same behavior again. The net result is that the team becomes more cohesive and committed to the project.

2. **Write letters of appreciation for team members:** It is the project manager's responsibility to write letters of appreciation for high-performing team members. Some organizations allow the team member to submit the recommendation for a cash award.

3. **Write letters of appreciation for functional managers:** This task is politically motivated. Writing letters of appreciation to functional managers who have lent their people to your project is rewarding them for doing so. When you reward behavior, you increase the probability of getting that behavior again in a similar situation. The letter of appreciation will have a positive effect the next time you need to borrow a person from a functional manager.

4. **Release the team members:** The project manager may release the team verbally but another way is to include the release date in the letter of appreciation for the team member. Give each team member a copy at the team party and send a copy to the team member's supervisor.

5. **Celebrate:** Every project should conclude with a party. The cost, time, activities, and place must be acceptable and within the norms of the organization. A catered lunch is one way. Pizza and beer at a local restaurant is another. The project manager should pay for the celebration if the organization will not. Paying for a few pizzas and beers will not cause any lasting damage to the project manager's financial status and it shows class!

Glossary

Activity: *See* Work package; Task.

Activity on the node chart: *See* Network diagram.

Actual time: An estimate of the time it will actually take to perform a work package. Actual time = effort time ÷ productivity. *See* Effort time; Duration time.

Ahead/behind schedule: Equation for determining the number of days or weeks the project is ahead or behind schedule. *See* Equation 13.6.

Analogy estimate: A top-down method that provides an approximation of the project cost that is useful during the initiation phase. The estimate from this method is not accurate enough to serve as the project budget. *See* Bottom-up estimate; Parametric estimate.

Assumption: A condition associated with the performance of a project that will have an impact upon the planning or execution of the project. The customer will provide physical security for the building materials and equipment for the project is an example. Assumptions are discussed in the initiation phase of the project life cycle.

Average effort time: In the PERT approach, average effort time is the weighted average effort time. See Equation 5.1.

Average reserve: The average amount of money estimated to be needed to deal with risks within the project. This amount is sometimes defined as the expected value of all the threats minus the expected value of all the opportunities. *See* activity 6, Chapter 9.

BAC: Original project budget (OPB); *See* Budget at completion.

Backward pass: A procedure whereby the late start and late finish of each work package are determined. The rules for the backward pass are (1) always subtract; (2) put the numbers in the lower corners; (3) go in the direction opposite to the arrows; and (4) when you reach a convergence, always keep the smaller or smallest number. *See* Figure 6.7.

Baseline: The customers' or sponsors' expectations relative to the project total cost, scope, and duration time constitute the terminal baselines. The project manager sets instrumental baselines for each work package with which to control the project to achieve the terminal baselines. *See* activity 1, Chapter 10.

Benefit to cost ratio: An economic viability approach where the anticipated financial return (benefit) from the proposed project is divided by the anticipated cost of the project. A benefit to cost ratio of 1.2 indicates a benefit of $1.20 for every dollar invested. *See* Net present value; Payback period; Simple profit.

Bottom-up estimate: An accurate technique for estimating the cost of a project: (1) develop a list of work package titles, (2) estimate the cost of each work package, and (3) add up the estimates.

Budget: The authorized amount of money to perform a work package or a project.

Budget at completion (BAC): An earned value term that refers to the original budget for the entire budget. *See* Original project budget (OPB).

Business case definition: The process of defining the numerous facets of a potential project to generate the information with which upper management makes an informed decision about funding the project.

Change control board: A group of people separate from the project team whose purpose is to assess objectively the feasibility and impact of scope changes. The project manager may or may not be a member.

Change control log: A document that lists all suggested scope changes and their final disposition. A part of the scope change procedure documentation. *See* Figure 11.4.

Charter: Document sometimes used to announce the official start of a project. It signals the start of the planning phase. Content varies but usually includes the mission of the project, identifies the project manager, and identifies the role, responsibility, and authority of the project manager. It may specify the procedures and documentation required of the project.

Client closeout: That part of the closeout plan that determines that all deliverables have been accepted and the degree to which the client is satisfied. *See* Final project report; Organizational closeout; Project closeout; Risk management closeout; Subcontractor closeout; Team closeout.

Closeout phase: Last of the four project life-cycle phases. Purposes of the closeout phase are to close the project in a disciplined way and to satisfy management that all aspects of the project have been resolved. Also called termination phase. *See also* Execution phase; Initiation phase; Planning phase.

Closeout plan: That part of the project plan that addresses those work packages that must be performed at the end of the project to close the project in a disciplined way. It includes client closeout, organizational closeout, project closeout, risk management closeout, subcontractor closeout, and team closeout. *See* Chapter 12.

Communications plan: Lists major stakeholders to the project and information each stakeholder wants or needs, and indicates the method that will be used to meet each stakeholder's needs. The purpose of the plan is to gain acceptance and support for the project.

Constraint: A condition associated with the performance of a project that limits the team's flexibility to perform the project. Not being able to install a system in a client's facility during the customer's normal work hours is a constraint.

Contingency reserve: A set-aside of resources time, money, people, or equipment with which to deal with risks that have been identified within project. *See* Management reserve; Reserve cost.

Control: The result of numerous project manager actions that seek to keep scope, cost, and duration variances within acceptable limits. *See* Control limits.

Control limits: There are two kinds of control limits: limits on the maximum and minimum duration for each work package, and limits on the maximum and minimum cost for each work package. The project manager sets the control limits for all work packages. *See* Lower control limit; Upper control limit.

Convergence: A point in the network diagram where two or more arrow points meet on the forward pass, or where two or more arrow tails (where the feathers would be if they were real arrows) meet on the backward pass. *See* Figures 6.6 and 6.7.

Cost: *See* Direct cost; Indirect cost; Project performance cost; Reserve cost.

Cost at 90% confidence level: An estimate of cost (either for a work package or the total project) that we are confident would be exceeded 10 or less times if we were to run the project 100 times; the estimate would not be exceeded 90 of the 100 times. Cost at 90% confidence level equals the effort time at 90% confidence level times the loaded labor rate of the person performing the work. *See* Figure 5.1.

Cost at 95% confidence level: An estimate of cost (either for a work package or the total project) that we are confident would be exceeded five or less times if we were to run the project 100 times; the estimate would not be exceeded 95 of the 100 times. Cost at 95% confidence level equals the effort time at 95% confidence level times the loaded labor rate of the person performing the work. *See* Figure 5.1.

Cost performance index (CPI): The efficiency with which money is being used on the project. A CPI of 1.2 means that the project has generated $1.20 of value for every dollar spent. *See* Equation 13.4 (determining the CPI); Equation 10.12 (forecasting final project cost from the CPI). *See* Schedule performance index.

Cost saving: An economic viability approach where the project is expected to generate a financial savings equal to the anticipated project savings minus anticipated project costs. This approach is useful where the internal project will save the organization money rather than generate revenue. *See also* Benefit to cost ratio, payback period, and simple profit.

Cost variance: The difference between what we planned to spend and what we actually spent. *See* Equation 13.2. *See* Schedule variance.

CPI: *See* Cost performance index.

Critical path: That path in the network diagram where work packages have zero total float (reserve of time). In the network diagram, the critical path is the path with the longest cumulative duration time. The duration of the critical path determines the project duration time.

Critical path SPI: The SPI for those work packages on the critical path. *See* Equation 10.10. It is the SPI on the critical path that determines the project's total duration time.

Customer satisfaction survey: A survey instrument to assess the degree to which the customer is satisfied with the project team's performance in meeting the customer's needs. *See* Figure 12.3.

Date number notation system: System whereby the early start, early finish, late start, and late finish of a work package are represented by calendar dates. *See* Day number notation system.

Day number notation system: System of numbers on the network diagram that indicate the early start, early finish, late start, and late finish days. The project starts on day number 1. Day numbers can be converted to calendar dates; see Table 6.1.

Definitive estimate: *See* Bottom-up estimate.

Deliverable: A tangible or intangible output of a work package or a project. Training-needs assessment is the tangible deliverable for the work package "Develop Training Needs Assessment"; the intangible deliverable for the work package "Conduct Regression Analysis Training" is students capable of performing regression analysis.

Dependency relationship: *See* Logic.

Direct cost: Project costs directly related to performing the work of the project and easy to calculate. Sometimes a direct cost is recaptured as an indirect cost because it is difficult to determine the amount of the cost; e.g., the cost of electrical energy consumed in the project room.

Duration time: An estimate of how long the project manager will wait to get the work package. Differs from effort time and actual time if the work package worker is working less than full time on the work package. Duration time = effort time ÷ availability. *See* Actual time; Effort time.

EAC: Estimate at completion. *See* Latest estimate of project cost (LEPC).

Early schedule: The numbers in the upper corners of the work packages (on the network diagram) that indicate the earliest day a work package may start (upper left corner) and the earliest day a work package may complete (upper right corner). The early schedule is determined by the forward pass. *See* Late schedule.

Earned value (EV): For a single work package or the total project, the dollar value of work completed. *See* Equation 13.1.

Earned value technology: A set of simple equations for assessing the progress made by a work package or the total project. It includes equations for forecasting final project cost and duration time. *See* activity 6, Chapter 13.

Economic viability: A generic term that refers to any number of economic equations, which address the financial return expected from the project investment. Economic viability is determined using the following techniques: benefit to cost ratio, simple profit, net present value (multiple-year profit), payback period, return on sales, return on investment, and economic value added. *See* Chapter 2.

Effort time: The original estimate of the time to perform a work package. Effort time is affected by performance to determine actual time. Effort time is affected by availability to determine duration time. *See* Actual time; Duration time.

Effort time at 90% confidence level: An estimate of effort time (either for a work package or the total project) that we are confident would be exceeded 10 or less times if we were to run the project 100 times; the estimate would not be exceeded 90 of the 100 times. *See* Equations 6.8 and 6.9.

Effort time at 95% confidence level: An estimate of effort time (either for a work package or the total project) that we are confident would be exceeded five or less times if we were to run the project 100 times; the estimate would not be exceeded 95 of the 100 times. *See* Equations 6.6 and 6.7.

Estimate at completion (EAC): *See* Latest estimate of project cost (LEPC).

EV: *See* Earned value and Expected value.

Execution phase: The third of the four project life-cycle phases. Purposes of the execution phase are to control the execution of the project, which includes delegating, controlling, and monitoring work; dealing with scope changes; and evaluating and forecasting project final cost and duration. Also called implementation phase. *See also* Closeout phase; Initiation phase; Planning phase.

Expected value (EV): An estimate of the overall effect (or benefit) of a threat (or opportunity). *See* Equation 9.3. Sometimes called expected monetary value (EMV).

Expected value of exposure: Probability multiplied by impact (in dollars or time) is the expected value of the exposure. Expected value of the exposure is the average impact (in dollars or time) of recovering from the threat if the project was run many times and the average impact determined. *See* activity 4, Chapter 9.

Expected value of leverage: Probability multiplied by impact (in dollars or time) is the expected value of the leverage. Expected value of the leverage is the average benefit (in dollars or time) captured from the opportunity if the project was run many times and the average benefit determined. *See* activity 4, Chapter 9.

Final project report: That part of the closeout plan that seeks to inform upper management of the completion of the project and to assure these people that all aspects of the project have been properly closed. *See* Client closeout; Organizational closeout; Project closeout; Risk management closeout; Subcontractor closeout; and Team closeout.

Final risk assessment: *See* Risk management closeout.

Finish-to-finish relationship: One of three possible relationships between two adjacent work packages on the network diagram. In the absence of a number on the arrow between the work packages, finish-to-finish means that the predecessor and the successor must finish on the same day. *See* Finish-to-start relationship; Start-to-start relationship; Lag; Lead.

Finish-to-start relationship: One of three possible relationships between two adjacent work packages on the network diagram. In the absence of a number on the arrow between the work packages, finish-to-start means that the predecessor must finish before the successor can start. *See* Finish-to-finish relationship; Start-to-start relationship; Lag; Lead.

Forecast of project final cost: *See* Latest estimate of project cost (LEPC).

Forecast of project final duration time: *See* Latest estimate of project duration (LEPD).

Forward pass: Procedure whereby the early start and early finish day numbers for each work package are determined. The four rules for the forward pass are (1) always add; (2) put the numbers in the upper corners of the work package; (3) always follow the direction of the arrows; and (4) when you reach a convergence, always keep the larger or largest number. *See* Figure 6.6.

Free float: A reserve of time that exists at the end of a network diagram path because the work packages on that path have not used their total float. Usually free float will equal the amount of unused total float from the preceding path. *See* Equations 6.11 and 6.12.

Gantt chart: A diagram showing the start and finish of all work packages on the project. It may show the planned and actual schedules. Named after its inventor, Henry Gantt.

Go-away estimate: An unfortunate attitude in some organizations where employees believe that accurately estimating the time and cost to perform a work package is an annoying interruption of their legitimate work. The estimator takes as little time and effort as possible to "pull an estimate from the air," and presents the estimate in a manner that implies "now go away!" to the requestor. This attitude should not be tolerated by the project manager.

Hunch technique: An undesirable method of justifying a project where some powerful individual has a hunch about the need for a project or the potential success of a project. Usually includes the launching of the project without further investigation or justification.

Implementation phase: See Execution phase.

In control: *See* Project performance assessment; Work package assessment.

Independent risk events: Refers to two outcomes from risk events that have no relationship to or impact upon each other. For example, the chance of John pulling a king in a single-card draw from a deck of cards at one table is independent of the chance that Mary will pull a club in a single-card draw at another table. The two outcomes are independent of each other.

Indirect cost: A cost not directly related to the performance of the project, or directly related to the project but difficult to determine. Indirect cost includes maintenance on the building, amortization and depreciation of equipment or buildings, upper management salaries (usually), and the costs associated with R&D, marketing, human resources management, etc. *See* Direct cost.

Initiation phase: First of the four project life-cycle phases. Purposes of the initiation phase are to determine the needs for the project, define the business case for the project, and get management approval of the project. Also called the concept phase. *See also* Closeout phase; Execution phase; Planning phase.

Instrumental baseline: The expected scope, cost, and duration time for each work package that is set by the project manager. If work package managers can perform their work packages within acceptable variances of their baselines, the final project will be achieved within acceptable variances of the terminal baselines. *See* terminal baseline.

Interrelationship: *See* Logic.

Labor cost: Equals the numbers of hours worked times the worker's loaded labor rate (LLR). For example, a worker with an LLR of $30 per hour works 40 hours on a work package; the labor cost equals $30 \times 40 = \$1,200$. *See* Loaded labor rate (LRR).

Lag: A positive number on the arrow to a successor work package (in a network diagram) that delays the start of that successor work package. *See* Figure 6.8 and Lead.

Late schedule: The numbers in the lower corners of the work packages (on the network diagram) that indicate the latest day a work package may start (lower left corner) and the latest day a work package may complete (lower right corner). The late schedule is determined by the backward pass. *See* early schedule.

Latest estimate of project cost (LEPC): The forecast of the project final total cost. It uses the cost performance index to forecast cost. An alternate term is estimate at completion (EAC). *See* Equation 10.12.

Latest estimate of project duration (LEPD): The latest estimate of project duration is based upon the schedule performance index along the critical path. *See* Equation 10.11.

Lead: A negative number on the arrow to a successor work package (in a network diagram) that causes the successor work package to start earlier. *See* Figure 6.9 and Lag.

Lesson learned: Experience that may be used to continue successful performance or to improve unsuccessful performance. *See* Figure 12.6.

LEPC: *See* Latest estimate of project cost.

LEPD: *See* Latest estimate of project duration.

Life cycle of product or system: Refers to the total life of a system, including the time it takes to develop (the project life cycle), deploy, maintain and repair, and dismantle and remove (scrap) the system.

Life cycle of project: Refers to the total life of a project from its inception through its completion and termination. Commonly broken into initiation, planning, execution, and closeout phases.

Loaded labor rate (LLR): The cost per hour for a person's time. Includes the direct labor cost, the indirect, and overhead costs.

Logic: Refers to the chronological order of work packages as shown on the network diagram. Other terms that mean chronological order are dependency, inter-relationship, and precedence.

Lower control limit: Refers to the smallest schedule or cost performance index that is acceptable. For example, the lower control limits for SPI and CPI are $LCL_{spi} = .90$ and $LCL_{cpi} = .95$, meaning that the SPI for all work packages must not fall below .90 and the CPI for all work packages must not fall below .95. The project manager establishes the upper and lower control limits. *See* Equation 10.2 (CPI); Equation 10.4 (SPI).

Management reserve: A set-aside of resources, usually money, with which to deal with risks that either have not been identified within the project or that impact the entire organization. The management reserve is held at the top of the organization; usually it is not available to the project manager.

Milestone: A significant event determined by the customer or the project team with a date on which it is expected to be completed.

Milestone chart: A chart showing the planned and actual dates for all milestones. Milestones may be shown on a separate document or incorporated into the Gantt chart.

Monte Carlo: A software package that uses the PERT three-point estimates to generate either a probability distribution of total project cost or a probability distribution of project total duration time (completion dates).

Mutually exclusive events: Events from the same risk where one or the other may occur, but not both. If you get a head on a coin toss, it means you did not get the tail; it is not possible to get both on a single toss. *See* Independent risk events.

Net present value: An economic viability approach where the anticipated profit from a multiple-year project is equal to the present value of the revenues minus the present value of the costs. *See also* Benefit to cost ratio; Payback period; Simple profit.

Network diagram: A graphic display of the chronological order of the project work. The network diagram determines the project total duration time and the critical path. Also called precedent chart and activity on the node chart.

Normal work days: Refers to the days the organization is normally open for work and the number of hours of work expected per day. For most organizations, the normal work days include Monday through Friday and eight hours per day. Normal work days vary with occupational groups. Security personnel work a seven-day week and each security person may work four 12-hour days. Also called working time.

OEPC: Original estimate of project cost. *See* Original project budget.

One-minus rule: The total probability of two mutually exclusive events (A and B) must equal one. If you know the probability of one of these events, you can determine the probability of the other by subtracting the known probability from one. *See* Equation 9.2.

OPB: *See* Original project budget.

Opportunity: One of the two kinds of risk events that may occur on a project. Opportunities produce positive impact: increases in revenue or shortening of the schedule. *See* Threat.

Opportunity accept strategy: Refers either to active acceptance where a plan is developed before the opportunity occurs to capture the benefit of the opportunity, or passive acceptance where no plan is developed before the opportunity occurs. The team develops a strategy after the opportunity has occurred. *See* Opportunity enhance strategy; Opportunity ignore strategy; Opportunity share strategy.

Opportunity enhance strategy: Refers either to taking a pre-emptive action to increase the probability of the opportunity occurring, or taking a pre-emptive action to increase the amount of benefit that accompanies the opportunity, or both. *See* Opportunity accept strategy; Opportunity ignore strategy; Opportunity share strategy.

Opportunity ignore strategy: Deciding not to pursue the opportunity before or after it occurs. It means the organization is not interested in the opportunity. *See* Opportunity accept strategy; Opportunity enhance strategy; and Opportunity share strategy.

Opportunity share strategy: A strategy whereby the team seeks out a third party to achieve the opportunity because the opportunity lies outside the team's expertise. *See* Opportunity accept strategy; Opportunity enhance strategy; and Opportunity ignore strategy.

Opportunity strategies: Strategies with purposes to accept, enhance, ignore, or share the benefits of opportunity events. *See* Threat strategies.

Organizational closeout: That part of the closeout plan that concludes the team's use of organizational resources and makes a final resolution of the remaining resources. *See* Client closeout; Final project report; Project closeout; Risk management closeout; Subcontractor closeout; Team closeout.

Original project budget (OPB): The total estimated cost to complete all the work packages in the project. It does not include the cost (reserve) for dealing with risk events. Also called OEPC and BAC.

Out of control: *See* Project performance assessment; Work package assessment.

Paired comparison: One of a number of techniques that rank a list of threats or opportunities or other items by taking the total list of items and comparing each combination of two at a time to generate the information with which to rank the items. *See* Table 2.3.

Parametric estimate: A top-down method that provides an approximation of the project cost that is useful during the initiation phase. The estimate from this method is not accurate enough to serve as the project budget. This method uses a single equation to estimate the cost of the entire project, e.g., cost of a house equals $65 per square foot multiplied by total square feet of the house. *See* Bottom-up estimate; Analogy estimate.

Payback period: An economic viability approach where the anticipated time required for a proposed project to generate the profits to repay the entire up-front investment is the criterion. Payback period may be in weeks, months, or years. *See* Equation 2.5 and 2.6. *See also* Benefit to cost ratio; Net present value; Simple profit.

PERT: *See* Program Evaluation Review Technique.

Planning phase: The second of the four project-planning phases. The purpose of the planning phase is to develop the planning documents with which to execute the project. Also called the development phase. *See also* Closeout phase; Execution phase; Initiation phase.

Precedence: *See* Logic.

Precedent chart: *See* Network diagram.

Predecessor work package: A work package that is performed before the successor work package. *See* Successor work package.

Present value of money: The concept that a dollar today is worth more than a dollar in the future because we can earn interest on money we have now. Present value is an equation that allows us to convert any amount of future money into its present value. Converting future dollar amounts to present value allows us to compare the relative cost and revenues from proposed projects. *See* Equation 2.3.

Probability: The likelihood of a risk event happening. *See* Equation 9.1.

Problem: A threat event that has occurred. *See* Windfall.

Productivity: The rate at which work is performed. Productivity is automatically included in the estimate of how long a worker will need to complete a work package as long as the original estimate (called effort time) is based upon that worker doing the work. Productivity becomes an issue when the person who performs the work package is different from the person around whom the original estimate was determined. *See* Equations 6.1 and 6.2.

Program Evaluation Review Technique (PERT): A series of equations used to estimate the average effort time of a work package, the standard deviation associated with the average, and other equations for adding a reserve of time or money into the estimates. PERT estimate of effort time is based on three estimates: longest time to complete (P), shortest time to complete (O_p), and the most likely (ML) time to complete. *See* Equations 5.1 and 5.2.

Project: A temporary effort by a team of people to produce some specific end result.

Project closeout: That part of the closeout plan that assures that the project is complete and that all requirements have been met. *See* Client closeout; Final project report; Organizational closeout; Risk management closeout; Subcontractor closeout; Team closeout.

Project dimensions: Dimensions that define the project include scope, time, and cost.

Project management: A disciplined approach to initiating, planning, executing, controlling, and terminating the activities of a team in its efforts to achieve an end result.

Project manager: The person responsible for leading the team through the activities required of the project.

Project performance assessment: Performed by comparing the CPI and the SPI for total project against the upper and lower control limits established by the project manager. Total project cost and schedule variance are said to be within acceptable limits when their corresponding indices (CPI and SPI) fall within the control limits. A project is "out of control" when the schedule or cost variance are beyond acceptable limits; this occurs when the corresponding CPI or SPI are outside the control limits. *See* Lower control limits; Upper control limits.

Project performance cost: The cost of performing all the work packages in the project. The estimate of this amount is called the original estimate of project cost (OEPC); this is the original budget for the project.

Project schedule: The documentation that indicates the start and end of all work on the project. The schedule is indicated on the network diagram and the Gantt chart.

Rating sheet: A subjective technique that uses a predetermined list of rating criteria with which a group of people rank alternative project proposals. The technique does not include any objective measures of economic viability.

Recipe: A step-by-step procedure to produce an end product or result.

Reserve: A set-aside of resources (money, people, equipment, time) with which to deal with risk events.

Reserve cost: The amount of money set aside to deal with risks (threats and opportunities) in the project. Sometimes called a contingency reserve.

Resource availability: (a) Refers to the total number of people scheduled to work on the project and whether they will be available in the numbers required by the plan. (b) Refers to the percent of time a person will work on a particular work package, e.g., Mary can complete the work package in 40 hours if she works full time on it but the project manager will have to wait 80 hours if she works on it half time (50%). *See* Equations 6.4 and 6.5.

Resource plan: A spreadsheet that indicates all the resources required to perform each work package of the project. Typically includes who will perform the work package, the start and end dates of the work package, the budget, the equipment, the material, and other special information relevant to the work package.

The resource plan is used by the project manager to manage the numerous activities of the project. *See* Table 7.1.

Risk: An future event that may have a positive effect (opportunity) or a negative impact (threat) upon the project. Risks are properly defined when we know or can estimate the event itself, the probability of the event happening, and the impact of the event.

Risk identification: Any of a number of techniques that identify threats and opportunities to the project. Techniques to identify risks include records analysis, expert interviews, surveys (Delphi), brainstorming, nominal group technique, and Crawford slip.

Risk management closeout: That part of the closeout plan that seeks to identify those threats and opportunities that are germane to the close of the project. Also called final risk assessment. *See* Client closeout; Organizational closeout; Project closeout; Subcontractor closeout; Team closeout.

Risk management plan: Plan that identifies threats and opportunities, estimates the probability and impact of each, and determines strategies for dealing with each threat and opportunity. The purpose of the plan is to maximize the benefits of the potential opportunities and minimize the impact of the potential threats.

Schedule performance index (SPI): Efficiency with which the project spends time. An SPI of .80 means that the project achieves 80% of a day's planned work for each day consumed. *See* Critical path SPI; Equation 13.5 (determining the SPI); Equation 10.10 (critical path SPI); Equation 10.11 (forecasting final project duration).

Schedule variance: The difference between what we planned to accomplish and what we actually accomplished. *See* Cost variance.

Scope: The total of all the work that must be performed on the project. Many documents discuss scope but the definitive scope document is the work breakdown structure (WBS).

Scope change log: Form used to summarize the initiation and final disposition of all scope changes to the project. *See* Figure 11.4.

Scope change procedure: A six-step process that controls, documents, and communicates scope changes to the project. *See* Chapter 11.

Scope change request form: Form used to document the initiation, analysis, and final disposition of each scope change. *See* Figure 11.1.

Simple profit: An economic viability approach where the anticipated net return from a proposed project (profit) is equal to the anticipated revenue minus the anticipated costs. Simple profit is applied to projects that are one year or less in duration because the time value of money is not a factor. *See* Equation 2.2. *See also* Benefit to cost ratio; Net present value.

Single-point estimate: A single estimate of how long the work package will take to complete; this is opposed by the PERT technique of estimating effort time, which is based on three estimates: longest time to complete, shortest time to complete, and the most likely time to complete. *See* Program Evaluation Review Technique (PERT).

Small-group techniques: A number of facilitated small group processes that can be used to generate ideas from a group of knowledgeable people. Includes brainstorming, nominal group technique, and Crawford slip. *See* Chapter 9, activity 2.

SPI: *See* Schedule performance index.

Standard deviation: A number that is a measure of the confidence you should have in the estimated mean effort time. Standard deviation provides an easy method to add a time or money reserve to the project with which to deal with schedule risk. *See* Equation 5.2.

Standard deviation of the critical path: One method of adding a reserve to a project schedule requires the determination of the standard deviation of the critical path. *See* Equation 9.4.

Start-to-start relationship: One of three possible relationships between two adjacent work packages on the network diagram. In the absence of a number on the arrow between the work packages, start-to-start means that the predecessor and the successor must start on the same day. *See* Finish-to-finish relationship; finish-to-start relationship; Lag; Lead.

Statement of work: A statement in the procurement solicitation and the contract that briefly outlines the purpose of the contract (project). Usually not definitive enough to serve as the scope document for performing the project; the work breakdown structure (WBS) serves this purpose.

Subcontractor closeout: That part of the closeout plan that determines that all subcontractors have completed their work and makes a final resolution of monies due. *See* Client closeout; Final project report; Organizational closeout; Project closeout; Risk management closeout; Team closeout.

Successor work package: A work package that is performed after the predecessor work package. *See* Predecessor work package.

Task: *See* Work package.

Team closeout: That part of the closeout plan that concludes the team's existence as a separate entity. *See* Client closeout; Final project report; Organizational closeout; Project closeout; Risk management closeout.

Terminal baseline: The customer's or sponsor's expectations of scope, time, and cost for the project at the conclusion of the project. *See* Instrumental baseline.

Termination phase: *See* Closeout phase.

Threat: One of the two kinds of risk events that may occur on a project. Threats produce negative impact, increases in cost or delay in schedule. *See* Opportunity.

Threat acceptance strategy: Refers either to active acceptance where a plan is developed before the threat occurs, or passive acceptance where no plan is developed before the threat occurs. The team develops a strategy to recover from the threat after the threat has occurred. *See* Threat avoidance strategy; Threat mitigation strategy; Threat transfer strategy.

Threat avoidance strategy: Refers to a strategy of completely eliminating the probability of the threat occurring by changing the circumstances or method to which the threat is attached. *See* Threat acceptance strategy; Threat mitigation strategy; Threat transfer strategy.

Threat mitigation strategy: Either of two preemptive strategies to (1) reduce the probability of a threat occurring or (2) to reduce the amount of impact (damage) that the threat produces. *See also* Threat acceptance strategy; Threat transfer strategy; Threat avoid strategy.

Threat strategies: Strategies that purpose to accept, avoid, mitigate, or transfer the impacts of threat events. *See* Opportunity strategies.

Threat transfer strategy: Refers to a strategy whereby the team finds some other person or organization to pay the impact of the threat. Collision insurance is an example of threat transfer because the cost of the collision would be born by the insurance company. *See* Threat acceptance strategy; Threat avoidance strategy; Threat mitigation strategy.

Top-down estimate: Either of two relatively inaccurate methods of estimating project cost, which provide estimates useful during the initiation phase when a "ballpark" approximation is appropriate. *See* Analogy estimate; Parametric estimate.

Total float: A reserve of time that work packages not on the critical path have that may be used either to delay the start of the work package, or to increase the duration time, or both, as long as delaying the start and increasing the duration do not consume more reserve time (total float) than exists. *See* Equation 6.10. *See* Free float.

Total project performance budget: *See* Original project budget (OPB).

Upper control limit: Refers to the largest schedule or cost performance index that is acceptable. For example, the upper control limits for SPI and CPI are $UCL_{spi} = 1.10$ and $UCL_{cpi} = 1.05$, meaning that the SPI for all work packages must not be greater than 1.10 and the CPI for all work packages must not be greater than 1.05. The project manager establishes the upper and lower control limits. *See* Equation 10.5 (SPI); Equation 10.3 (CPI). *See* Project performance assessment; Work package assessment.

Variance: (a) A statistical term referring to the square of the standard deviation. Standard deviation is three days; the variance is nine days. (b) Plan minus actual. Applied to both cost and time. If the plan calls for a total cost of $100,000 and actual cost is $120,000, variance is –$20,000 ($20,000 over budget). If the plan calls for completing the project in 80 days and it is completed in 70 days, variance equals 10 days (10 days early). *See* Cost variance; Schedule variance.

WBS: *See* Work breakdown structure.

WBS dictionary: *See* work package work order.

Windfall: An opportunity that has occurred. *See* Problem.

Work breakdown structure (WBS): An organized list of all the work that must be performed on the project. It lists the work by work package titles and the work packages are listed under categories of work. *See* Tables 4.1 through 4.8.

Work package: The smallest unit of work listed on the work breakdown structure (WBS). The level at which work is delegated, monitored, and controlled by the project manager. "Task" and "activity" are synonyms, although "task" is usually used for informal projects such as organizing a picnic, and "activity" and "work package" are usually used on formal projects like building a ship.

Work package assessment: Performed by comparing the CPI and the SPI for every work package against the upper and lower control limits established by the project manager. Work package cost and schedule variance are said to be within acceptable limits when their corresponding indices (CPI and SPI) fall within the control limits. A work package is "out of control" when the schedule or cost variance are beyond acceptable variance; this occurs when the corresponding CPI or SPI are outside the control limits. *See* activity 2, Chapter 10.

Work package work order: A form that contains all the information needed by the work package manager. Sometimes called WBS dictionary. *See* Figure 4.1.

Working time: *See* Normal work days, the preferred term.

Workload leveling problem: Places in the project schedule where there is too much work scheduled and/or not enough work scheduled for the number of people on the team. *See* workload leveling solution.

Workload leveling solution: A strategy of shifting work within its float to ameliorate the workload leveling problem. *See* activity 2, Chapter 7.

Appendix B

Compact Disk

Table of Contents

Chapter 2 Initiation Phase

Chapter 4 Project Scope

Chapter 5 Project Cost

Chapter 6 Project Time (Schedule)
Table 6.2 Network Diagram Checklist

Chapter 7 Project Resource Plan
Table 7.1 Template for the Project Resource Plan

Chapter 8 Project Filing System
Table 8.1 List of project files

Chapter 9 Risk Management
Table 9.1 Format for the threat list
Table 9.2 Format for the opportunity list
Table 9.3 Interview outline
Table 9.6 Risk cost estimating sheet
Figure 9.6 Spreadsheet that calculates the time-reserve for the critical path

Chapter 10 Project Baseline and Control
Table 10.2 Manual template for a variation control matrix
Figure 10.2 Spreadsheet template for the variation control matrix
Figure 10.4 Spreadsheet template for project performance report

Chapter 11 Project Scope Change
Figure 11.1 Scope change request form with embedded procedure
Figure 11.4 Scope change log form

Chapter 12 Project Closeout Plan
Figure 12.1 Final project evaluation outline
Figure 12.2 Outline for project final report or briefing
Figure 12.3 Customer satisfaction survey
Figure 12.5 Outline for final risk assessment

Chapter 13 Execution Phase
Figure 4.1 Work package work order

Index

T - #0116 - 101024 - C0 - 234/156/16 [18] - CB - 9781138440432 - Gloss Lamination